FIXING THE CLIMATE

T0002718

Fixing the Climate

Strategies for an Uncertain World

Charles F. Sabel

David G. Victor

PRINCETON UNIVERSITY PRESS

PRINCETON AND OXFORD

Copyright © 2022 by Princeton University Press

Princeton University Press is committed to the protection of copyright and the intellectual property our authors entrust to us. Copyright promotes the progress and integrity of knowledge. Thank you for supporting free speech and the global exchange of ideas by purchasing an authorized edition of this book. If you wish to reproduce or distribute any part of it in any form, please obtain permission.

Requests for permission to reproduce material from this work should be sent to permissions@press.princeton.edu

Published by Princeton University Press
41 William Street, Princeton, New Jersey 08540
99 Banbury Road, Oxford OX2 6JX

press.princeton.edu

All Rights Reserved

First paperback printing, 2024
Paperback ISBN 9780691224534

The Library of Congress has cataloged the cloth edition as follows:
Names: Sabel, Charles F., 1947– author. | Victor, David G., author.
Title: Fixing the climate : strategies for an uncertain world /
Charles F. Sabel, David G. Victor.
Description: Princeton : Princeton University Press, [2022] |
Includes bibliographical references and index.
Identifiers: LCCN 2021041684 (print) | LCCN 2021041685 (ebook) |
ISBN 9780691224558 (hardback) | ISBN 9780691224541 (ebook)
Subjects: LCSH: Environmental policy—International cooperation. |
Climatic changes—International cooperation. | BISAC: POLITICAL SCIENCE /
Public Policy / Environmental Policy | LAW / Environmental
Classification: LCC GE170 .S24 2022 (print) | LCC GE170 (ebook) |
DDC 363.738/74526—dc23/eng/2022128
LC record available at https://lccn.loc.gov/2021041684
LC ebook record available at https://lccn.loc.gov/2021041685

British Library Cataloging-in-Publication Data is available

Editorial: Bridget Flannery-McCoy and Alena Chekanov
Production Editorial: Sara Lerner and Nathan Carr
Jacket/Cover Design: Michel Vrana
Production: Erin Suydam
Publicity: Kate Hensley, James Schneider, and Kathryn Stevens
Copyeditor: Cindy Milstein

Jacket/Cover Credit: iStock

This book has been composed in Adobe Text and Gotham

From Charles F. Sabel:

To my daughter, Francesca, who showed me how to build back, better

From David G. Victor:

For my children, Apple and Eero, an inspiration to get serious about healing the planet

CONTENTS

ACKNOWLEDGMENTS

This book emerged over seven years, starting with an invitation from Bob Keohane, Gráinne de Búrca, and Rick Locke to a November 2014 seminar at Brown University on experimentalist governance. Once we had full drafts of the manuscript, we benefited in particular from detailed comments from Tom Hale and two anonymous reviewers at Princeton University Press.

Early on, we presented a draft of the major ideas at seminars at Princeton University, Columbia University, Indiana University, and the Paris Institute of Political Sciences. Chuck is grateful to Piero Ghezzi, Ron Gilson, Bernard Hoekman, Jeremy Kessler, Rory O'Donnell, Dani Rodrik, Robert Scott, and Jonathan Zeitlin for collaborations, in parallel, that helped sharpen the ideas. David thanks Simon Sharpe, Frank Geels, and Danny Cullenward for their collaborations that helped evolve concepts along with their application to climate policy.

Our analysis of the Montreal Protocol in chapter 2 benefited from discussions with Ted Parson, and our application of those lessons to climate change in that chapter (and chapter 6) was sharpened by Dan Bodansky—especially where Dan disagreed with us. The discussion of deliberation in chapter 3 is indebted to Dan Ho's work and his comments on our interpretation of it. The case studies in chapters 4 and 5 relied heavily on expert inputs. Rory O'Donnell and Larry O'Connell, the past and current directors of the National Economic and Social Council, were partners in developing the Irish water case, and in relation to the California Air Resources Board case, we depended on the extraordinary research that Lauren Packard did for a seminar paper at Yale Law School. Our case study on US sulfur regulation was aided by insights from Tom Alley, Kerry Bowers, Tony Facchiano, Nanda Srinivasan, Arshad Mansoor, Randall Rush, Tom Wilson, and especially Chuck Dene, who walked us through the early years of sulfur control technology. For the Advanced Research Projects Agency–Energy case, we learned from Erica Fuchs, Michael Piore, Arun Majumdar, Laura Diaz Anadon, Kelly Sims Gallagher, David Hart, and Ellen Williams. Our research on the Brazilian Amazon was guided by Stephan Schwartzman, Daniel Barcelos Vargas, Salo Coslovsky, Pablo Pacheco, Roberto Mangabeira Unger, and invaluable research by Vitor Martins Dias and Gustavo Fontana Pedrollo as well as commentary on our case study by the distinguished participants in an online seminar organized by Daniel Vargas. Our case study on integrating renewables into the California grid benefited from insights from Jeff Dagle, Reiko Kerr, Ben Wender, Tom Overbye, Sue Tierney, Ken Silver, Larsh Johnson, and particularly Josh Gerber and Rachel McMahon.

Near the end of this project, Josh Cohen and Deb Chasman invited us to excerpt the book for debates in the December 2020 issue of the *Boston Review*; the debates, which occurred just as the United States was contemplating a possible Green New Deal that could have evolved in experimentalist ways, helped us push this book to the end. It also introduced us to Matt Lord, who edited our piece in the *Boston Review* and became our editor for the book as a whole. Almost everyone needs an editor who, like Matt, combines the rigor of the mathematician and the intellectual fearlessness of the philosopher. We certainly did. The peerless Jenny Mansbridge read the book at the very end with her usual acuity and insight. Her influence on our thinking began long before that; it was, in keeping with the book's theme, atmospheric and helped encourage our inchoate efforts to connect novel responses to uncertainty with the reimagination of democracy.

Special thanks to Linda Wong and Jackson Salovaara for exceptional research assistance, Évita Yumul, Steve Carlson, Emily Carlton, Jen Potvin, and Kate Garber for their help with the references, and again, and above all, Évita, whose technical prowess, unmatched resourcefulness, and generous commitment to the project saved us again and again from the consequences of our computer-assisted bumbling. The Stanley Foundation, The Brookings Institution, Columbia Law School, Electric Power Research Institute, Norwegian Research Foundation, and the University of California at San Diego all provided financial support. At Princeton University Press, thanks to Eric Crahan, David Campbell, Alena Chekanov, Cindy Milstein, and especially the patient Bridget Flannery-McCoy for steering this study to publication, and helping us muster the courage of our convictions throughout.

LIST OF ABBREVIATIONS

APA Administrative Procedure Act
ARPA-E Advanced Research Projects Agency–Energy
CAISO California Independent System Operator
CAR Cadastro Ambiental Rural
CARB California Air Resources Board
CDM Clean Development Mechanism
CFC chlorofluorocarbon
COP Conference of the Parties
CO_2 carbon dioxide
CPUC California Public Utilities Commission
DARPA Defense Advanced Projects Agency
DOE Department of Energy
EHC electrically heated catalyst
EPA Environmental Protection Agency
EPRI Electric Power Research Institute
ETS emission trading system
ICLEI Local Governments for Sustainability
IPCC Intergovernmental Panel on Climate Change
LAWPRO Local Authority Water Programme
LEV low-emission vehicle
MLF Multilateral Fund
NAAQS National Ambient Air Quality Standards
NARUC National Association of Regulatory Utility Commissioners
NDC nationally determined contribution

NEPA National Environmental Protection Agency (China)
NGO nongovernmental organization
NO_x nitrogen oxide
NRDC National Resources Defense Council
ODS ozone-depleting substances
OPA open plurilateral agreement
PSD prevention of significant deterioration
PT Workers' Party (Brazil)
PTA preferential trade agreement
PZEV partial-zero-emission vehicle
REDD+ Reducing Emissions from Deforestation and Forest Degradation
SO_2 sulfur dioxide
TEAP Technology and Economic Assessment Panel
TOC Technical Options Committee
UN United Nations
UNEP United Nations Environment Programme
UNFCCC United Nations Framework Convention on Climate Change
WFD Water Framework Directive
WTO World Trade Organization
ZEV zero-emission vehicle

FIXING THE CLIMATE

1

Introduction

TOWARD EXPERIMENTALIST GOVERNANCE

Can the world meet the challenge of climate change?

After more than three decades of global negotiations, the prognosis looks bleak. The most ambitious diplomatic efforts have focused on a series of virtually global agreements such as the Kyoto Protocol of 1997 and Paris Agreement of 2015. But with so many diverse interests across so many countries, it has been hard to get global agreement simply on the need for action, and *meaningful* consensus has been even more elusive. Uncertainty about which emissions reduction strategies work best has impeded more robust action; prudent negotiators have delayed making commitments and agreed only to treaties that continue business as usual by a more palatable name. All the while, emissions have risen by nearly two-thirds since 1990, and they keep climbing—except for the temporary drop when the global economy imploded under the coronavirus pandemic. Yet to stop the rise in global temperature, emissions must be cut deeply—essentially to zero over the long term.

Meanwhile, similar problems have plagued global governance more generally. The World Trade Organization (WTO), founded in 1995, has been paralyzed for more than a decade by the kind of consensus decision-making that has hamstrung climate diplomacy. In many other domains, from human rights to investment to monetary coordination, international order seems to be fraying. With no global hegemon and no trusted technocracy—welcome changes in the eyes of many—there is no global authority to mend it.

Popular protest has only reinforced this global gridlock. The Great Recession of 2008 exposed the limits of the postwar model of economic growth, and the economic shock triggered by the pandemic has dramatically exacerbated social inequality. No wonder that climate change and economic policy have become even more densely intertwined politically. For conservatives in many countries, decarbonization is a fraught symbol of the global elite. Repudiating climate agreements—Donald Trump's snubbing of the Paris Agreement, for example—has been seized on as a way to reassert the primacy of national interests after decades of unchecked globalism. For progressives, meanwhile, efforts to reconcile sustainability and inclusive well-being find expression in calls for massive public investments such as a Green New Deal. That vision has found tentative success in only a small fraction of the global economy—one that accounts for a shrinking slice of global emissions.

But bleak as it is, this record is not the whole story. Alongside the string of disappointing global agreements and false visions of surefire solutions are significant as well as promising successes in many other domains. We can learn from them in the fight to rein in warming. From the global to the local levels, and at every level in between, models of effective problem-solving have already emerged and continue to make progress on issues, like climate change, that are marked by a diffuse commitment to action, but no clear plan for how to proceed. These efforts work in countries as diverse as China, Brazil, and the United States, and for international problems as diverse as protecting the ozone layer and cutting marine pollution. They address challenges as intrusive and contentious as any that arise with deep decarbonization, and tackle challenges whose solutions require unseating powerful interests and transforming whole industries. In sector after sector, from steel to automobile transport to electric power, real progress in the elimination of emissions is gaining momentum.

The strategy underlying these initiatives points the way forward. They work by setting bold goals that mark the direction of the desired change. But they acknowledge up front the likelihood of false starts, given the fact that the best course of action is unknowable at the outset. They encourage ground-level initiative by creating incentives for actors with detailed knowledge of mitigation problems to innovate and then converting the solutions into standards for all. But they also enable ground-level participation in decision-making to ensure that general measures are accountably contextualized to local needs. When experiments succeed, they provide the information and practical examples needed to mold politics and investment differently—away from vested interests and toward clean development. They solve global problems not principally with diplomacy but instead by creating new facts on the ground—new industries and interest groups that benefit from effective problem-solving, and that push for further policy effort.

We call this approach to climate change cooperation *experimentalist governance.* It is sharply at odds with most diplomatic efforts—including the important but ultimately flawed Paris Agreement—which so far have failed to make a meaningful dent in global warming. The architects of global climate treaties assumed that the dangers of climate were clear, and that solutions were in hand or easily discoverable. The real problem—in their understanding, often the only one—was the allocation of the costs of adjustment and the associated mobilization of political will. Since cutting emissions is expensive, and each nation is tempted to shirk its responsibilities and shift the costs to others, climate diplomats took it for granted that no nation would cooperate unless all are bound by the same commitments. The analogy was to a group of shepherds, aware that together they are overgrazing the commons they share, but each calculating that it is foolish to reduce their flocks unless all the others do. From those assumptions came the requirement that climate change agreements should be global in scope and legally binding. The result is global action no more ambitious than what the least ambitious party will allow.

These assumptions have not stood up to the test of time, and neither has the paradigm for solving the climate problem. Above all, the easy availability of solutions can't be taken for granted. The experience of recent decades with, for example, electric vehicles, integration of renewables in the power grid, and improvements in ground-level pollution control, shows the difficulties. While solutions can be achieved, they are hard to come by and require deep, coordinated changes in many domains. Progress depends on the degree to which innovation is encouraged and coordinated. From this perspective, the problem that the overgrazing shepherds face is not primarily to agree on sharing the burdens of adjustment but to make adjustment feasible by cooperating to develop a new breed of sheep that grazes on less grass—and perhaps new varieties of grass and pasture practices as well. If that metaphor captures the fundamental challenge of climate change, then the best way to build effective consensus is not to ask who will commit to certain predetermined outcomes no matter what but instead to begin by systematically encouraging solving problems at many scales and piecing the results together into ever-stronger solutions. Global commitments, achieved through diplomacy, should be the outcome of our efforts rather than the starting point.

This is a book about extraordinary but little-noticed innovations in organization and governance that take this alternative approach. We show

how experimentalist strategies work under conditions of deep and pervasive uncertainty about the right solutions even when familiar approaches fail. We illustrate how they link local action with more encompassing coordination to speed the solution of general problems and, conversely, how they adapt general solutions to local contexts. We explain how public, private, and civil society actors, monitoring themselves and each other, can work together to advance decarbonization while making the economy more efficient and nimble. Along the way, we revisit enough of the history of climate change agreements to explain how the dominant institutions of the day all but foreclosed effective cooperation. Our central aim is to reorient our current climate change regime away from failed efforts based on ex ante global consensus, and toward a system anchored in local and sectoral experimentalism and learning. We firmly believe we can meet the stark challenges before us, and experimentalist governance shows us how.

A paradigm case of experimentalist governance and central example running through this book is the 1987 Montreal Protocol on Substances That Deplete the Ozone Layer—by many measures, the single most effective agreement on international environmental protection. We argue that we still have a lot to learn from Montreal as well as a lot to *unlearn* from mistaken views about the basis of its accomplishments. To set the stage for the rest of the book, we give a preview in the following section of the nuts and bolts of the protocol's exemplary successes. We then spell out the fundamental principles that made it work: the bedrock design ideas of experimentalism, which we will explore in more depth in later chapters. Next, we identify three flaws of traditional climate change policy thinking that impede more effective forms of action and go on to discuss how all of this relates to the signature piece of climate diplomacy today: the Paris Agreement of 2015. We end with the plan for the rest of the book.

The Montreal Protocol: An Exemplary Success

Crafted in the late 1980s, the Montreal Protocol was ahead of its time.[1] Not only was it highly effective, but it became a model for what might be achievable in solving the problem of climate change. Despite widespread admiration for the successes of Montreal, the real reasons for its achievements were largely misunderstood and misapplied in the case of climate change. Although the ozone and climate regimes looked quite similar on the surface, Montreal advanced quickly to solve the ozone problem while there was little problem-solving in the domain of climate change.

It is useful to go back in history to probe why Montreal worked—and how it became an exemplary system of experimentalist governance. That proper understanding is essential to knowing not just why the ozone layer is healing but also how to make more progress on climate change by creating an institutional architecture that takes uncertainty for granted—a system that is a spur to innovation rather than a cause of political gridlock.

Beginning in the 1970s, scientists detected chemical reactions thinning the atmospheric ozone layer that protects most life on earth from ultraviolet radiation. The cause was traced to the emissions of chlorofluorocarbons (and later other chemicals, including halons) that were then widely contained or used in the manufacture of many products, from aerosol sprays to fire extinguishers, styrofoam, refrigeration and industrial lubricants, and cleaning solvents. After more than a decade of contentious debate, two linked treaties, the Vienna Convention (1985) and Montreal Protocol (1987), created the framework for a global regime whose governance procedures were elaborated in the following years. The original black letter provisions in these agreements were thin on content; success came from how these institutions evolved through practice. Nobody used the term "experimentalist governance" to describe what they were doing, but experimentalism is the system that they created.

The core of this system of governance is a schedule to control and eventually eliminate nearly all ozone-depleting substances (ODS). The measures are reassessed every few years in light of current scientific, environmental, technical, and economic information, and the schedule is adapted as necessary. The periodic meeting of the parties has broad authority to review the implementation of the overall agreement, and make formal decisions to add controlled substances or adjust schedules.

In this regime, problem-solving is broken down into sectors that use similar technologies, and is guided by committees representing industry, academia, and government regulators. The committees organize working groups of ODS users and producers to review and assess efforts, mainly in industry, to find acceptable alternatives. The reviews consider key individual components as well as whole systems—for example, assessing whether a refrigerant that depletes the ozone layer can be replaced by an analogous and more benign alternative as well as whether refrigeration systems that utilize these new chemicals can work reliably and at an acceptable cost. Pilot projects yield promising leads that attract further experimentation at a larger scale, allowing the committees to judge if the nascent solution is robust enough for general use. Without the institutions of the Montreal

Protocol, what looks like the successful spontaneous search for alternative technologies would not have been possible.

If this search comes up short, the committees and their oversight bodies authorize exemptions for "essential" and "critical" uses, or extend timetables for phaseout. When the use of ODS was phased out in the metered dose inhalers that propel medication into the lungs of asthmatics, for instance, the sectoral committee consulted doctors, pharmaceutical companies, and device manufacturers country by country to determine substitutes along with transition schedules that met the safety and efficacy requirements of patients. When a few firms invented an array of alternative metered dose inhalers using benign propellants, the committees put the industry on notice that the old methods would be banned. Innovative firms had a strong incentive not to be left out, and persistent laggards faced exclusion from the market.

Over time, an amendment procedure allowed additions within the existing categories of coverage and also brought new categories of emissions under control. The boundaries around "sector" were adjusted as the properties of each class of ODS was understood and new sectors were implicated. Analysts often celebrate Montreal because it followed the science of ozone depletion, but that science at the time of Montreal's adoption was indeterminate as to ozone safe solutions, and the real root of success was the Montreal orchestration of experimentation and learning about uncertain industrial futures.

Membership in the Montreal Protocol expanded sharply as well. Initially the protocol focused on industrialized countries, as they had the highest consumption of ODS and were most compelled politically to stop ozone thinning. But use increased rapidly among developing countries, and they were allowed to extend their compliance schedules so as to encourage their participation in the protocol. As a further incentive, essentially all the costs of compliance for developing countries were paid by the Multilateral Fund (MLF) financed by the rich countries—costs that included not just the new technologies but also the local administrative capacity needed to oversee the preparation and execution of comprehensive regulatory plans for phasing out the production and use of ozone-destroying chemicals sector by sector. Simply making new technology available would not have compelled the use of these benign alternatives; local contextualization was essential, and the fund helped build that capacity. Administratively, the fund is probably the best-managed funding mechanism in the history of international environmental governance. Politically, it helped transform the ozone problem from one with a guaranteed deadlock—since developing countries did not want to bear all of these costs themselves—into one that was more practical politically.

The Montreal regime operates against the backdrop of vague but potentially draconian penalties for governments and firms that drag their feet. For the Western governments that initiated the regime, such as the United States, those penalties were electoral. (Those were the bygone days when the United States was a reliable leader on global environmental topics.) For the industrial firms that made the noxious substances, the penalties were about brand value and the license to operate. DuPont, the most visible of these firms and therefore the most vulnerable, broke ranks with the rest of the industry to demand a phaseout. (It helped that the alternatives might prove more profitable.) Once there was one innovator, it was too costly for others to lag behind. And in countries that actively undermine the Montreal Protocol—Russia at first, but others later on, including India and China—the penalties were threats such as trade sanctions that came from other powerful governments, mainly in the industrialized world, that wanted Montreal to work and also wanted to make sure their home industries would not be undercut by violators overseas.

Designing for Uncertainty

The features of the Montreal approach that make it a good model can be captured in a handful of design principles. Together they characterize a distinctive decision-making process that is well suited to domains, like climate change, marked by great complexity and uncertainty where the very nature of possible outcomes is unknowable in advance.

This approach starts with a thin consensus among an open group of founding participants motivated to act. The precise definition of problems, let alone the best way to respond to them, can't be anticipated at the outset, but there is enough agreement on how to get started. In the case of Montreal, that initial agreement took the form of an acknowledgment that ozone thinning was a problem that must be stopped, and a first step would require cutting in half the most widely used ODS by 1998. At the time there was no agreement on the magnitude of the risk, the feasibility of finding particular substitutes by certain dates, or even whether 50 percent cuts were the right goal. Consensus thickens with effort, however, and new knowledge demonstrates what is needed, and which actors are capable and trustworthy. Interests are mutable as actors come to anticipate an advantage in the destabilization of the status quo and more demanding regulation. Participation is open, in the sense that new actors outside the circle of founders are invited in as their experience and expertise become relevant to addressing core problems.

In this scheme, the actual problem-solving is devolved to local or front-line actors—those most likely to have the kind of experience and expertise that embodies unanticipated possibility and unsuspected difficulty. Under Montreal, the most essential ground-level work has been technological, and performed by industrial enterprises developing and testing new chemicals and equipment along with local regulators that figure out how this equipment will operate in real-world conditions—for example, how metered dose inhalers can meet drug safety standards.

This local problem-solving is regularly monitored by a more comprehensive body. In the case of Montreal, assessment panels and sectoral committees periodically take stock of local problem-solving and help codify lessons. Monitoring is typically implemented by peer review: actors with overlapping but distinct areas of expertise and experience evaluate particular projects against others of their kind. The fund monitors projects in developing countries, and updates pooled knowledge about what actions cost and whether they work—vital information because each time Montreal parties adjusted or amended regulatory obligations, they also needed to update the funding plan. These routines help spot and scale successful innovation, and make it easier to nip budding failures. Just as an initial, broad understanding of problems is corrected by local knowledge, so local choices are corrected in light of related experience elsewhere.

A comprehensive review leads, in turn, to periodic adjustments along with a redirection of means and ends. From a distance, Montreal looks like a regime that always ratcheted commitments tighter, but viewed close-up, it becomes apparent that progress was less linear. Goals were periodically relaxed through exemptions and deadline extensions when problems proved unexpectedly hard. Science helped identify broad goals, but the pace of on-the-ground problem-solving—along with what the parties were willing to spend through the MLF and other funding mechanisms—determined compliance deadlines and the timing of additions to the list of regulated substances. Periodically, a centralized assessment panel takes stock of the lessons, and offers a plan for how emission controls could be adjusted, the benefits to the ozone layer, and what it would cost.

A distinctive combination of penalties and rewards incentivizes both public and private participation in this type of regime. By rewarding leaders to bet on change, they make it risky for laggard firms and government to bet against it. This *penalty default*, as it is known, destabilizes the status quo;

obstruction becomes the riskiest bet of all. And once the logjam of current interests is broken, shifting the question from *whether* change is possible to *how* it can be implemented in diverse conditions, the failure to keep pace is viewed more as a symptom of ignorance and incapacity than as an expression of selfish cunning.

The initial form this feedback effect takes is to call attention to shortfalls and offer assistance, not punish wrongdoing. Only when misbehavior persists and comes to seem incorrigible does the reaction become draconian: actors that repeatedly prove unwilling or unable to improve are threatened with expulsion from the community, typically by being excluded from key markets.

These principles are unfamiliar in the realms of climate policy because much of that world frames climate change correctly as a problem of global collective action, but incorrectly equates global problem-solving with the search for solutions through consensus diplomacy. Most diplomacy, we will suggest, largely follows and aids on-the-ground experimentation and problem-solving rather than leading from the front. These principles, however, are not alien to the regulators, firms, and nongovernmental organizations (NGOs) that have stumbled onto ways of working together to solve hard problems. They have discovered that the only way to move beyond the status quo is to destabilize it, and then learn, quickly, to use the daring and imagination that bubble up in the open space to develop better approaches.

Experimentalist Governance Hidden in Plain Sight

This experience of managing under conditions of complexity and uncertainty is familiar to regulators and firms working on ground-level problem-solving. To understand why it has not translated easily into international efforts, it is helpful to take a closer look at conventional assumptions. In particular, policy choices have often been structured around three false dichotomies.

The first and most consequential is the view that organizations are either top-down or bottom-up. Top-down organizations are bureaucracies of the kind we associate with big corporations or big government. Precise goals are set at the top, and translated into detailed rules or operating routines in order to direct execution. Frontline workers apply the rules or follow the routines; middle managers see that they do or make ad hoc adjustments as necessary. Bottom-up organizations, for their part, seem hardly like organizations at all; they are forms of coordination that emerge as actors—ideally on equal footing, left to themselves, and given enough time to suffer the consequences of their mistakes—eventually master common problems.[2]

The Paris meeting was a victim of this top-down, bottom-up dichotomy. It was convened in the recognition that top-down climate organization, culminating in the Kyoto Protocol, had failed. The parties to Paris took that failure to mean one had to embrace bottom-up organization. But the opposite of a failure does not make a success. Bottom-up organization under real-world conditions—where some actors are much more powerful than others, local agreement is often perturbed by outsiders, and time for decisions is short—is merely a recipe for churning and inaction without direction and discipline.

By contrast, experimentalist governance is neither top-down, like a hierarchy, nor bottom-up, like a self-organizing group. It is both in turn, as lower levels of institutions correct higher ones and vice versa. Mindful that climate change actors are too heterogeneous in their interests and capacities for self-organization, experimentalist governance imposes top-down framework goals and penalty defaults to give direction to bottom-up invention. It provides incentives to both capable, potential innovators and less capable, potential laggards to encourage advances that are ultimately workable for all. This combination of seemingly incompatible features makes experimentalism especially suited to areas like climate change that carry a significant degree of uncertainty.

The second and closely related false dichotomy that has hindered progress on climate change is the choice between technocracy and democracy. In this vision, organizations are either hierarchically controlled by technocrats and managers asserting or pretending to expertise, or else they are democratically accountable to their members and other stakeholders.

One of those who saw past this dichotomy was the pragmatist philosopher John Dewey. Dewey took uncertainty and change as the dominant problems of political life, and the need to adapt institutions to new circumstances as the continuing challenge to democracy. The response, he argued, was to explicitly acknowledge the fallibility of current arrangements, and make concrete problems the trigger to the adjustment of methods and clarification of goals. But he cautioned that the collaborative investigation of alternatives can only be effective if it integrates the knowledge of experts with the experience and values of citizens, for it is the citizen who knows best "where [the shoe] pinches, even if the expert shoemaker is the best judge of how the trouble is to be remedied."[3] The broad participation of stakeholders in the Montreal sectoral committees provides a glimpse of how such cooperation can work. As trust in elites frays in our democracies and decarbonization reaches deeper into everyday life, this kind of working collaboration between shoemakers and shod is increasingly important. It is how systems of governance—even at the international level—will earn and retain greater democratic accountability.

A third misleading dichotomy pits organizations against markets. Decision-making in organizations is said to be centralized, initially by command—or when rules run out, by discussion and deliberation. In markets, by contrast, decision-making is supposed to be decentralized, with coordination achieved by prices.

This distinction has proved in its own way as limiting as the top-down and bottom-up dichotomy thanks to its application to thinking about carbon markets. Though they were not included formally in the founding agreements on climate change, carbon markets quickly became integral to the ideal conception of a global regime. As soon as emissions reductions targets were set, it became clear that their very rigidity entailed the need for some compensating flexibility. Market mechanisms seemed to square the circle, such as cap-and-trade schemes and offsets that allowed polluters with high costs of abatement to buy permits to pollute from those who have low costs of control. As individual actors minimize the costs of or returns from abatement, the overall effect is a gain in what economists call "static efficiency."[4]

The really big gains in pollution reduction, however, come not from the optimization of current practices but instead from destabilizing innovation—innovation that sharply reduces the carbon footprint of a product or whole production process, or even completely redefines an entire industry. Achieving these transformative outcomes is difficult. Producing the next generation of familiar technology is relatively straightforward and cheap; striking out in radically new directions to create much cleaner technology is risky and expensive in comparison. The rewards and penalties needed to directly incentivize that shift would have to be high and speculative—so high and so speculative as to make them politically unacceptable. For these and other reasons, pure market instruments have never imposed limits severe enough or prices high enough to test the effects of high-powered incentives on innovation. There is scant evidence that in their normal operation, they contribute much to "dynamic efficiency"—efficiency over the longer term, as technology and interests are changing.

Experimentalist governance, we will argue, makes a start at filling in this oversight in the discussion as well. Experimentalist institutions straddle the dichotomy between markets and organizations. They encourage and build on the kinds of decentralized or localized individual initiative and coordination we associate with markets.[5] But unlike markets, local decisions in experimentalist systems don't influence each other merely through prices. Rather, they also and often most directly ramify through processes like standards setting and revision that depend on discussion and deliberation—discursive processes, as in organizations.[6] In fact, we will contend that especially with regard to dynamic efficiency—on which progress toward a sustainable world ultimately depends—it is by this combination of price incentives and discursive, decentralized coordination that experimentalist governance can make good on the promise of carbon markets.

Finally, experimentalism agrees with the literature on international regime complexes in marking the demise of consensus-based, hierarchical, and global governance institutions.[7] In that vacuum—a gridlock of governance—the regime complex literature has documented the rise in many domains of disjointed constellations and partial regimes, pursuing sometimes complementary and sometimes conflicting purposes, and in the absence of any superior authority, forced to negotiate relations among themselves.[8] The literature on regime complexes focuses in fact on describing the emerging processes of negotiations, and the distributions of role and authority that may result from them. Experimentalist governance concentrates instead on the way regime complexes—metaregimes—can provide the context for experimentalist organizations, most especially in the crucial case of the Paris Agreement. For decades, scholars have viewed the climate change problem as one that requires giant, global contracts, with parties facing strong incentives to breach.[9] By contrast, we see cooperation emerging from the process of learning through experiments and the adjustment of interests in tandem.

Beyond Paris

How could all of this redirect climate policy strategies today? Experimentalist governance, we argue, provides a set of tested principles to guide the construction of regimes that do a good job of managing problems steeped in uncertainty when conventional organizations can't. There remains a role for international diplomacy, such as under the Paris Agreement, but that role is considerably smaller than its enthusiasts think. Successful problem-solving requires experimentation—a process that occurs mainly within countries and industrial sectors, not orchestrated through global agreements. While there is a role for a more centralized review and assessment of that decentralized experimental information, one of the lessons from the Montreal experience concerns how to design and evolve the institutions useful for that review and assessment. In the main, however, the way forward is to work sector by sector, within institutions that have the ability to apply experimentalist governance.

With regard to warming-related emissions in particular, it is useful to distinguish two types of sectors.

At one extreme are sectors comprised of globalized and highly concentrated industries, such as aircraft, steel, cement, auto, gas, and oil, whose products or production methods are subject to international standards. In these sectors, deep decarbonization entails risky and costly *innovation* at the frontier of technology, often driven by penalty defaults. International cooperation is appealing to firms under these conditions because it allows them to pool in some measure knowledge and risks—the Swedish steelmaker bets on one radical alternative to the current methods, the American on another technology, and periodically they carefully compare notes—and by demonstrating the feasibility of alternatives, they can raise standards and protect themselves against cutthroat competition from firms that continue to produce the traditional way. Thus Maersk, the world's largest container shipping company by fleet size and cargo volume—and thus the firm best positioned to gain from successful advances—coordinated a series of technology demonstration programs inside the International Maritime Organization that was cofunded by governments and linked to proposals for new standards. Because cargo ships are long-lived and hard to change once built, Maersk also works with these same governments to gradually align equipment and local standards to superior solutions, proving the workability of many paths to improvement and making it easier for other International Maritime Organization members to join in.

At the opposite extreme are more place-based sectors such as residential and commercial construction and power grids incorporating clean energy sources. In these cases, production is largely for local markets, using many local inputs, even if key components like wind turbines, nuclear fuel, or flooring materials are global commodities. The central challenge for international cooperation at this extreme is not simply innovation but also *contextualization:* making new technology work reliably in various places, according to local circumstances. Standards that shape these industries are more likely to be local and national than international. Integrating renewables on California's grid is different than doing so on India's, even though both buy solar panels from the same global market. Cooperation can accelerate emissions reductions by pooling learning; even if solutions are quintessentially place based, they typically result from the reelaboration of innovative techniques developed elsewhere. Knowing where to start and what doesn't work under conditions similar to one's own is invaluable information.

In between these two extremes are hybrid cases, such as forestry products or palm oil, where the inputs are predominantly local, but the markets—and the standards and trade barriers that control access to them—are international. Reducing illegal logging or burning forests to clear land for agriculture requires reaching deep into local economies, often under limited control by the national state, to give small producers lucrative and stable alternatives to the current, environmentally destructive ones. Progress here is slow, but it continues.

This sectoral taxonomy matters because it informs where to focus effort, how to organize it, and where to look for the many signs of progress already emerging. The Paris Agreement can't guide, much less participate directly in, sectoral experimentation at the frontier of technological innovation or the contextualization of place-based solutions. However, there has been a profusion of problem-solving efforts along these lines within other forums. Some are informed by experimentalist principles. If anything, there is a surfeit of national and international organizations directed to these tasks. The challenge for international cooperation on climate change today isn't creating new sectoral institutions as much as identifying and coordinating the efforts of those that do or could work.[10]

Even though Paris has little to contribute directly to this process, it does serve one essential and exclusive function. It is the most legitimate institution in global politics where climate change is discussed; it sets goals that while probably impossible to meet, are widely agreed on as a starting point. In short, it is the climate conscience of the world. Its presence makes it easier for governments, firms, and NGOs to punish—in the name of Paris—actors that drag their feet. Without Paris, it would be much more challenging—politically and legally—for protesters to rattle companies that cause big emissions and push governments to act on climate change. These are the penalty defaults that destabilize the status quo and motivate innovation, and they are essential to our vision of experimentalist-driven decarbonization.

In fact, precisely because we see the fate of climate action as bound up with the development of other international organizations and efforts, this book is also about a broader transformation of the world order. A new climate change regime, evolved from the foundations of Paris but in more experimentalist directions, foretells a new kind of globalization—one, we argue, that is already in the works.

The Plan of the Book

In chapter 2, we trace the history of climate change diplomacy, attributing its failures to departures from the experimentalist lessons at the heart of the Montreal Protocol. Both the 1987 Montreal Protocol and 1992 United Nations Framework Convention on Climate Change (UNFCCC) were founded with similar, sparse legal language. What's different is how they evolved. With Montreal, which we examine in greater depth, the evolution turned it into a system for experimentalist governance that made it successful. Climate change evolved differently, with efforts overly focused on global diplomacy and the crafting of global consensus rather than experimentation and learning. We argue that these diplomatic efforts ultimately foundered, in part, because they drew the wrong lessons from Montreal. If Paris is to avoid remaking these and related mistakes, its architects will need to understand why experimentalism works and how it can be applied in practice.

In chapter 3, we present the theoretical underpinnings of experimentalism and illustrate its operation in practice. We look at the emergence of new forms of contract and administration that assume that the precise outcome of collaboration cannot be determined ex ante, and therefore that goals and methods have to be elaborated provisionally—step by step through experimentation across a wide range of opportunities, along with joint reviews of progress in which partners assess and come to rely on one another's capacities. In this setting we show why penalty defaults, in contrast to conventional fines for the infraction of clear rules, are the kind of sanctions appropriate to conditions of uncertainty. And we explain how institutionalized deliberation, often in the form of peer review, is essential to evaluating the lessons of experimentation, guiding further inquiry, and informing eventual standards. While this book is focused on climate change, the logic of experimentation applies to solving a wide range of problems marked by deep uncertainty.

The following two chapters look at experimentalist governance in action.

Chapter 4 examines experimentalist *innovation*. We explore three case studies of public-private collaboration at the technological and policy frontier: innovation in a range of key energy technologies by the Advanced Research Projects Agency–Energy (ARPA-E) of the US Department of Energy (DOE); the development of scrubbers to control sulfur dioxide (SO_2) pollution in the context of a pioneering cap-and-trade system of pollution permits; and the work of the California Air Resources Board (CARB) on vehicular emissions standards. As conventionally understood, these examples illustrate the three competing approaches to addressing climate change: one that looks to the state, one that looks to the market, and a third that splits the difference by leveraging the power of the state to threaten exclusion from the market. Rather than identify one of these approaches as optimal, we contend that in all of these modes of innovation, experimentalism is essential. The more profound and disruptive the innovation, the greater the need for institutions designed with a recognition that the right answers are unknowable ex ante, and only through experimentation via joint action between government and business is it possible to identify practical solutions. We deliberately take all three examples from the United States in order to repudiate the notion that successful public-private collaboration is culturally or politically impossible here.[11]

Chapter 5 explores the other prong of sector-based action, experimentalist *contextualization*: government-industry collaborations that turn technological advances into reliable on-the-ground systems in particular places. Again we consider three case studies: control of agricultural pollution in Ireland, the emerging regime to combat illegal logging in Brazil, and the integration of renewable energy into a power grid. We show how, in the absence of any overarching design, regulators, firms, farms, and NGOs are nevertheless creating—in all but name—expansive environmental protection regimes, stretching from the ground to the national or international level. Across all of these cases, central governance mechanisms help to establish which approaches are working in context and revise higher-level goals. Once those goals are (provisionally) set, rules and operating routines are contextualized to local circumstances.

In chapter 6, we apply the logic of experimentalism to international cooperation for deep decarbonization. In sectoral innovation at the frontier, the characteristic challenge for governance is conciliating progress by a small vanguard of innovators with the ultimate inclusion of the rest of the (initially less capable) global economy. In contextualization, the challenge is to accelerate and reduce the costs of reciprocal learning and capacity building among regions facing similar local problems—and using this capacity building to augment local political support so that progress on decarbonization is less vulnerable to changes in the political wind.[12] Much of the research on the politics of climate change has emphasized how organized interest groups block policy; with experimentalism, properly applied, we see a mechanism through which some of those groups find new interests, and politics becomes both dynamic and pointed toward decarbonization.[13] We find exemplary institutions engaged in innovation and contextualization; the immediate task for policy, we argue, is finding or building more, not asserting that Paris itself can perform these functions. Paris, though, does retain an essential role. As the climate conscience of the world, its legitimacy can be leveraged to induce groups to act in the name of Paris in establishing or applying penalty defaults to firms or governments.

Finally, in chapter 7, we look beyond climate change to the future of globalization more generally. Governing in the midst of uncertainty—while avoiding both gridlock and unaccountable technocratic control—is a generic problem of international affairs and solving shared global problems, not unique to climate change governance. The same troubles beset the coordination of trade policy and the WTO, the very core of the global economy. In this chapter, we call for reforms in trade cooperation—away from consensus-based, globe-spanning institutions, and toward the piecing together of expansive regimes from smaller, open, more collaborative, and accountable initiatives. Indeed we show that such transformations are already taking place, mirroring similar developments in climate change governance. Together these initiatives point the way toward a radical new form of globalization—one that advances piecemeal, by narrow agreements rather than all-or-nothing global commitments, and keeps action democratically accountable by remaining under sovereign control. Globalization should be reimagined, in the image of the climate regime's successes, to respect uncertainty and difference. This book is a tool for that reimagination.

2

Lessons from the Path Not Taken

MONTREAL AND KYOTO

Over a five-year period, from 1987 to 1992, governments signed international treaties on two of the most profound challenges in international environmental governance: the thinning ozone layer and warming climate. On the first front, the 1987 Montreal Protocol set an initial schedule for cutting the pollutants most harmful to ozone, authorized the creation of expert committees to evaluate further controls, and required governments to keep convening to consider next steps.[1] On the second front, the 1992 United Nations Framework Convention on Climate Change (UNFCCC) likewise set goals for cutting emissions, created committees to focus on technological and implementation challenges, and mandated subsequent meetings—again with the aim of progressively tightening controls on emissions.[2] Both treaties exempted developing countries from their obligations or granted them long delays in imposing any controls on their pollution.

On paper, the two documents were quite similar.[3] Indeed many of the same diplomats negotiated the two texts, and in the recollections of prominent participants, Montreal was the model for what came after in climate change.[4] But despite these connections, the institutional machinery created to implement the two agreements differed almost from the beginning. Over time these differences shaped the evolution of the agreements themselves, obscuring the original appearance of commonality.

The performance of the two treaties diverged sharply as well. The Montreal Protocol has manifestly led to deep reductions in emissions of ODS, to the point that the ozone layer is now healing itself, albeit slowly. By contrast, the UNFCCC control mechanisms—and the subsequent 1997 Kyoto Protocol, which turned the general obligations of the UNFCCC into specific and binding commitments—have had little appreciable effect on the emissions of warming gases.[5] While the UNFCCC agreements have doubtless contributed to a growing awareness of climate dangers, the failures of this apparatus are a significant part of the reason that the global pollution of greenhouse gases has been rising and planetary warming has been accelerating.

This chapter seeks to explain these divergences. The root cause, we maintain, is to be found in the underlying logic and institutions of governance in the two regimes. These design choices were, as is often the case with historical events of this degree of complexity and uncertainty, powerfully shaped though not compelled by differences in the background conditions under which the treaties arose.[6]

In the case of Montreal, we trace the incentives and background ideas that led to an experimentalist regime that drew firms and governments into collaboration in solving concrete pollution problems. We show how this form of problem-solving, once institutionalized, led almost immediately to an explosion of innovation that made it possible to set and implement ever-tighter limits on pollution. When progress did stall, the same institutional machinery allowed for the resetting of targets based on informed judgments about the level of pollution control that industry and governments could achieve. In time, the process of joint problem-solving and the results it produced led to deep changes in the interests of the actors. Major industrial firms previously opposed to pollution controls became enthusiastic supporters. Leading developing countries, from China to Brazil to Vietnam, shifted from resisting to embracing—sometimes zealously—pollution reduction efforts.

By contrast, we show how the UNFCCC regime—as it came to be embodied in the Kyoto Protocol—was not designed for collaborative, concrete problem-solving. Rather, it was designed to set unambiguous limits on allowable pollution, and then leave matters of implementation to national governments and firms on the logic that they alone knew the best ways to control pollution in their own circumstances. The diplomatic processes that arrived at those limits kept industry at a distance in the fear that public-private cooperation would invite regulatory capture. In the eyes of many of the architects of the UNFCCC agreements, this separation of responsibilities between diplomatic processes (controlled by government) and implementation (ultimately dominated by firms) could be taken one step further: given credible, country-specific binding limits on emissions, a market could be established that would allow the trading of quotas that would reward and give appropriate weight to superior solutions. Kyoto thus included several trading mechanisms that in theory, assured an efficient outcome by rewarding polluters with the lowest cost of abatement—regardless of sector, pollutant, or location—for intensifying their efforts. The reliance on market incentives as necessary and sufficient to induce change reflected the growing orthodoxy in the 1990s in the advanced countries and international economic governance that favored markets over bureaucracies as the instruments for solving problems.[7] The more complex the problem and unpredictable the solutions, this thinking had it, the greater the need to let market forces find the way. Yet in the case of climate change this approach produced essentially no innovation—certainly no burst of new technologies and strategies of the kind that accompanied the installation of the Montreal machinery—and arguably, no reduction of pollution beyond what participants would have undertaken in the absence of any agreement.[8] If anything, it produced a hardening of the initial conflicts of interest among the parties, deepening the division between advanced and developing countries. If diplomacy evolved into problem-solving with the development of the Montreal regime, in climate change, diplomacy remained in control.

This story has profound lessons for the value of experimentalist governance. But before we proceed to tell this story in detail, we have to address a common objection: that limiting climate change and protecting the ozone layer are such fundamentally different problems that methods effective against the one cannot be usefully applied to the other. On this view, ozone protection is the lesser challenge. It involved merely a few large, advanced countries and a few large, highly capable ODS-producing firms that stood to benefit if regulation drove established chemicals from the market. Moreover, this objection goes, the science connecting ODS to a dangerous deterioration of the ozone layer, harmful to human health and natural ecosystems, was clearly established almost from the start, and the technologies needed for safe substitutes were easily identified. In climate change, the opposite is true. The problem sprawls across virtually every aspect of life, involving countless actors with widely differing capacities. The science, though clear in its direction, remains daunting in its complexity, continuously generating differences among experts that critics can construe as confusion and doubt to avoid costly regulation, and the technologies needed for solutions sit far beyond the frontier of innovation. In short, the path to an ozone solution was obvious and easily traversed, while in climate it remains anything but.

There is something to this contrast. The climate problem implicates essentially all of industry and agriculture, while ozone-destroying chemicals are more marginal. For this reason alone, the reduction of warming pollution touches off more complex and ferocious political fights than disputes over ODS. There are more subtle differences as well. Ozone diplomacy came first, as developing countries were just beginning to articulate their interests in debates about how to govern the global environment. They were not much involved in the original crafting of the Montreal Protocol.[9] By contrast, in climate diplomacy, which began just a few years later, developing countries were much more aware that emission controls might slow their own economic development, much more alert to the injustice of being charged for climate harm caused by the accumulated pollution from advanced countries, and much better organized politically and diplomatically.[10]

But despite these differences, little else in the contrast of the two problems withstands scrutiny—and least of all the suggestion that the solution to the ozone problem was significantly more obvious and easier to implement. From the time the first alarms were sounded regarding ozone depletion, it took some fifteen years of public mobilization in various countries to create the setting for negotiating an international agreement, with many starts, stops, and wrong turns. The political and technological solutions to the problem were not obvious to the eventual parties to the ozone treaty. The science of ozone depletion likewise progressed, as does nearly every new subdiscipline of science, by fits and starts. The discovery of a "hole" in the ozone layer above Antarctica in winter 1985, right in the middle of the diplomatic process that would lead two years later to Montreal, revealed profound uncertainties in ozone science. Prevailing models of atmospheric science did not predict such a hole, much less whether human activity had caused it.[11] And while the science wavered, the technological prognosis was steadfastly discouraging. Leaving aside limits on the ODS in aerosols (where they were mostly used as propellants for hair spray and other toiletries, and easily replaced), a RAND study in 1980 predicted that it would be possible to find substitutes for at most 25 percent of ODS, at any price; a second authoritative study, reporting disagreements among the many experts consulted, allowed it could be possible to aim for, at most, a 50 percent reduction.[12] When the first international set of cuts was agreed on in Montreal, governments and industry were unsure what level of pollution control would be achievable.[13]

Only when the Montreal regime created sectoral committees—in which ODS-using firms were invited to join with government regulators and academic experts in searching for alternatives—did the full array of solutions start to emerge. Sectors such as industrial solvents and flame-quenching halons, where substitutes were widely presumed to be extremely costly or even nonexistent, quickly reduced or eliminated ODS use.

The lesson is clear. In the case of Montreal, it was not the confident expectation of quick and painless solutions that led to a problem-solving regime but rather the formation of such a regime that made the problem manageable. The problem of ozone depletion is certainly more circumscribed than that of climate change. But far from being categorically different, it is of the same general type, subject to the same kinds of uncertainty regarding scientific and technological understanding—and the same kinds of political disputes as well. What works for one is instructive about what can work for the other. Unfortunately, though, what worked in Montreal was not tried in the case of climate change.

We turn first to the story of Montreal, discussing its experimentalist institutions in detail. Then more briefly, we explore why, at nearly the same time, and in the absence of any credible threat of regulation or any encouragement of cooperation among firms and governments, the climate regime went in a sharply different direction, toward a system in which pollution reductions would be secured by the profit-maximizing decisions of individual firms instead of a collaborative search for new possibilities for greener production. Scholars and policy makers have learned a great deal from the failures of Kyoto in climate, although the efforts to apply those lessons remain fraught. We think there is just as much to learn from the successes of Montreal in protecting ozone, and this chapter aims to redress the balance.

The Origins of Environmental Diplomacy and the Success of Montreal

The prehistory of global environmental diplomacy stretches back to the end of World War II. The first wave of postwar international environmentalism led to international agreements on topics such as wetlands, endangered species, ocean dumping and regional air pollution, and in 1972, the creation of the United Nations Environment Programme (UNEP) as the "environmental conscience" of the international system.[14] UNEP rose in prominence through its work on projects such as an action plan to clean up the Mediterranean Sea—a plan that became a model for other regional seas around the world.

By the mid-1980s, concern focused increasingly on systemic, potentially irreversible assaults on the global environment such as deforestation, the thinning of the ozone layer, the reduction of biological diversity, and eventually greenhouse warming.[15] Mitigation in these cases required the provision of expensive public goods, with attendant collective action problems whose solution called for enhanced international cooperation.[16]

Even as the need for global cooperation increased, the terms of cooperation were changing. A growing number of developing countries were industrializing rapidly at the time, becoming or promising to become significant polluters in their own right.

Growth—and the sense of economic prowess that came with it—added new force and meaning to the Global South's postcolonial resentment of the West's hegemony in world politics. Developing countries demanded the same right to achieve prosperity that the advanced countries, in disregard of the environment, had claimed before them. From this followed the demand for compensation for contributions to global public goods along with a rejection of onerous commitments and intrusive monitoring that could inhibit growth. Managing the tensions between developed and developing countries, with the latter increasingly well organized and assertive, compounded the already-large challenges of managing global common pool resources just as ozone and then the climate emerged on the international agenda.[17]

It was under these conditions that atmospheric scientists discovered in the 1970s that the chlorine in CFCs depleted the ozone layer through catalysis. Each molecule of these chemicals could destroy many multiples of ozone, meaning that small quantities of industrial pollution could yield big global harms. Because CFCs are easily shifted from the liquid to gas phase, and are inert and nontoxic too, they were widely used as refrigerants as well as propellants in spray cans and for blowing plastic foams in industrial production. Some CFCs and related compounds are also good solvents that could be used safely as cleaning agents for degreasing industrial machinery or removing residues from circuit boards during soldering. Although most of the hypothesized ozone destruction came from chlorinated compounds, bromine—just below chlorine in the periodic table of the elements—was almost immediately implicated too. Brominated substances such as halon were widely used in fire extinguishers; other compounds of this class were used to fumigate agricultural products, among diverse other applications.

The discovery that the ozone layer might be in jeopardy mobilized public opinion in a number of advanced countries and led quickly in the United States to legislation with long-term consequences. Headlines in the US press announced the coming "Death to Ozone" and that "Aerosol Spray Cans May Hold Doomsday Threat." Environmental groups, rapidly expanding in size and influence, responded almost immediately: the Natural Resources Defense Council proposed a ban on CFC aerosols within months of publication of the paper linking chlorine to ozone depletion; public campaigns shaming consumer products companies for using dangerous substances resulted in a 25 percent decline in the sales of aerosols; several states banned aerosol CFCs in spray cans.[18]

Public pressure quickly reached Congress as well. The 1976 Toxic Substances Control Act defined "toxic" broadly enough that the Environmental Protection Agency (EPA) used it to prohibit CFC aerosol propellants two years later. More important, the 1977 amendments to the Clean Air Act, undertaken mainly to deal with local and regional air pollution, were extended to require the EPA to conduct regular assessments of the ozone problem, and instruct it to propose regulations for "any substance, process or activity" that may reasonably be expected to harm the ozone layer.[19] Acting on this provision, the EPA created a special office to consider the regulation of ozone thinners, making the eventual imposition of controls entirely likely.

Firms responded with a mixture of damage control and defense of the status quo. Users of ODS in aerosol packaging introduced a massive "CFC Free" labeling program to reassure the public that their spray cans did not threaten the environment. The large producers of ODS, meanwhile, challenged the new claims directly, arguing that the science was inconclusive, alternatives to CFCs were unsafe, and brash policy would result in "direct economic, health and safety" harms.[20] The CEO of DuPont, the biggest US producer of ODS and the company with the brand most at threat, tried to blunt criticism by pledging to cease production of CFCs if conclusive evidence linking them to atmospheric harm should be found—little anticipating that the company would have to honor his pledge in a few years.[21]

A few of the firms in ODS-facing industries undertook research programs, modest in relation to their budgets for public relations campaigns in this period, to explore the feasibility of substitutes. A few close molecular relatives of ODS showed promise, but synthesis at a commercial scale looked to be prohibitively expensive, and there were concerns with toxicity. DuPont abandoned its investigations in 1980. In retrospect the research programs were plainly defensive, designed to permit the firms to capture the gains from innovation should they prove easily achievable, but in the more likely case that progress proved elusive, the programs would provide evidence against mandating a costly regulatory leap into the unknown.[22]

Then there was a lull. Bans on some applications in the United States and a few other countries flattened global growth in the total use of CFCs for a few years. But increasing consumption in other countries and for uses not covered by the restrictions erased the gains. By the 1980s, the global consumption of ODS was on the rise again.[23] To further complicate the picture, the consensus in ozone science wobbled; as pollution increased, it seemed that ozone depletion, though still a serious problem, might not be as bad as originally feared.[24]

Against this backdrop of shifting science and intense yet sporadic public concern, UNEP convened meetings to discuss a treaty to address the problem.[25] Proceeding deliberately, but under little pressure of urgency, the diplomacy resulted in early 1985 in the Vienna Convention for the Protection of the Ozone Layer.[26] It contained few obligations except that governments should keep meeting and talking.[27] The governments most engaged in this diplomatic process were principally the Western industrialized nations under the greatest public pressure to address the ozone problem.

That same year, in 1985, as mentioned above, a hole in the ozone layer was detected exactly where and when climate scientists least expected it: over Antarctica, in the darkness of late winter. At the time, the chemistry of ozone depletion assumed that bright sunlight was required for the key reaction that broke apart chemical bonds and led, after many further steps, to ozone destruction. Dark places should have been the most protected from this process, allowing for the greatest accumulation of ozone. The Antarctic ozone hole was thus as much a shock for science as for industry and politics. Even though its causes were not initially understood, it showed the ozone problem could be much more urgent and mysterious than expected.

The negotiations for what would soon become the Montreal Protocol, underway even before the Vienna Convention entered into force, turned urgent. On September 16, 1987, diplomats in Montreal signed an agreement obligating the parties to cut CFC use in half within two years and freeze the use of halons. For most of Europe and Japan, reaching this target meant little more than applying the bans on CFC propellants already in place in the United States, while US compliance would require breaking new ground, in accord with the growing domestic pressure for action. By a coincidence so nice as to seem contrived, the science achieved important progress that very day: a converted U2 spy plane flown into the Antarctic hole with sensitive instruments measured a sharp increase in chlorine concentrations just as ozone concentrations dropped—data that analyzed later, confirmed the role of human agency in the form of CFCs causing the hole.[28] Soon thereafter the scientific puzzle was on the way to a solution; the catalysis of chlorine and bromine were indeed at work, but through a process not anticipated in the research of the early 1970s. The destructive chemistry in the Antarctic hole occurred on the surfaces of particles in the icy clouds of the polar stratosphere, without the need for much sunlight.[29]

For firms, too, the discovery of the ozone hole, in combination with the resumption of growth in the market for CFCs (and compounded by embarrassing revelations about their abandoned search for substitutes), tipped the scales from temporizing to active cooperation with regulators. Often the first to break rank were the users of ODS, which unlike the established producers of ODS, had no reason to prefer any particular substitute process over others. As they began cooperating with each other and the EPA in the search for alternatives, they also gave shape to the nascent institutions of the Montreal Protocol, extending this government-industry cooperation globally.

How Montreal Works

The success of Montreal depended on the way control measures were assessed and reviewed.[30] But interestingly, the treaty text governing that process—Article 6 of the 1987 Montreal Protocol—is skeletal.[31] It requires the parties to evaluate existing controls and the need for new ones at least every four years, beginning in 1990, "on the basis of available scientific, environmental, technical and economic information," and to convene for this purpose panels of experts in each of these broad areas at least one year before the reviews.

Notwithstanding this spare instruction—indeed, empowered in part by its vagueness—the experts from government, industry, and academia chairing the initial committees or writing chapters in their reports immediately proposed elaborations to the bare scheme in the text, and the parties accepted them. Their first, tentative efforts revealed that there was nothing like a blueprint for institution building at the start; a committee, no sooner called into existence, was fused with another or spawned subcommittees that became fundamental in their own right. But progress was rapid, and the purposiveness of the quick reorganizations indicated a shared willingness to undertake collaborative research on alternatives and a ready understanding of how to put that research into operation.

By the first Meeting of the Parties in 1989, a structure and working division of labor had emerged. The Scientific Assessment Panel was to report periodically on the overall health of the ozone layer and, if need be, propose extensions of the scope of the whole regime—the kind of comprehensive assessment of progress that would come to be called stocktaking in climate change discussions.[32] Concrete problem-solving was to be done in Technical Options Committees (TOCs), one for each sector, grouping industry specialists with corresponding regulators and researchers. Initially, TOCs were established in refrigerants, foams, solvents, aerosols, and halons; other TOCs were created as the regime expanded. Each TOC would become immersed in the process of implementation—for example, by identifying and helping to clear bottlenecks in industry, and linking suppliers of new technologies to users. The TOCs would also evaluate exemptions to commitments where elimination of a particular use proved premature, as well as the feasibility of accelerated phaseout schedules that allowed the parties, if they chose, to ratchet commitments tighter. The work of the TOCs was overseen by the Technology and Economic Assessment Panel (TEAP), which would look at the feasibility of adding new chemicals and uses for which a TOC wasn't yet in place.[33] Together the TOCs and TEAP organized the exploration and evaluation of promising alternatives in each sector, and translated the findings into goals that, while demanding, demonstrably met the test of feasibility.[34]

For the scientists and engineers organizing the new regime, there was nothing novel about the idea of collaboration in the solution of shared problems. As they were connected by education and work experience to professional communities that value reciprocity and honor elegant solutions, such cooperation was almost a reflex. For the concerned firms, however, the impulses conflicted. Having avoided engagement with regulators and, even more, one another in the preceding decade for fear of encouraging controls, firms had incentives to avoid collaboration. But as the likelihood of limits increased dramatically, the attraction of cooperation increased.

Two closely related developments helped tip the balance in favor of cooperation. First, the growing threat to US industry from Japan and other countries had sparked a national discussion about the need to foster precompetitive industrial research and development. The US Congress passed the National Cooperative Research Act in 1984 to encourage such activity, most notably by effectively exempting it from antitrust liability. Precompetitive research became a legitimating trope for organizing collective industry responses to novel challenges. Second, the EPA actively used the 1984 Act to advance the collaborative pursuit of ODS substitutes. Even before the Montreal structure had taken shape, the EPA had supported the formation of research consortia for the reduction of halons, speeding the deployment of alternatives in automotive air-conditioning, and crucially, testing the toxicity and environmental sustainability of new fluorocarbons. It then played a similar role within the ozone regime.[35]

Stephen Andersen, the head of industry cooperative programs in the EPA's Stratospheric Protection Branch as well as cochair of numerous Montreal committees and panels, including the Solvents TOC, was a key link between Montreal and the agency.[36] The scope of Andersen's activity illlustrates the fluidity of the emerging arrangements, degree of the EPA's commitment to joint efforts, and breadth of industry's disposition to cooperation. To the Solvents TOC, he recruited AT&T and Nortel, both of which used CFC-based solvents extensively in the assembly of circuit boards. Work on this TOC convinced both companies of the mounting advantages of collaboration, and the experience led them to form a separate international research consortium—the Industry Cooperative for Ozone Layer Protection, which would later play an important role in helping developing countries adopt ODS substitutes.[37] In addition to actively encouraging the formation of such consortia, the EPA sponsored annual conferences to circulate the latest developments in technology and alert industry to the direction and pace of regulation.[38] When standard setting or similar coordination problems stalled progress, the EPA convened ad hoc negotiations to restart it. Quickly the Montreal TOCs and other Montreal institutions began to coorganize such events with the EPA, and the line between national regulation and international institutions blurred. The EPA was reducing the costs of cooperation in the United States by helping potential partners find each other and supplying templates for organization, while at the same time increasing the costs of inaction with the promise of increasing regulatory controls. The TOCs, meanwhile, were propagating these actions internationally. The emerging cooperative equilibrium proved as compelling as had the compulsion of the status quo.

The Solvents TOC demonstrated the advantage of the joint search for solutions over the search capabilities of even the most capable incumbents left to themselves. Critically, it looked across the whole supply chain and beyond to other industries, not just at producers of the ODS. Established producers focused on the most familiar alternatives to the substances they already manufactured, assuming that the trajectory of development they had mastered would continue to be the path to superior solutions. Users were different. Technically sophisticated users such as AT&T and Nortel, which had a shared interest in finding an acceptable ODS substitute but no interest in producing that substitute themselves, were not attached to any particular solution, and thus were willing to search widely and together.

AT&T, for instance, had begun in the late 1970s to replace CFCs in circuit board assembly with aqueous solutions for reasons of cost as well as environmental and worker safety, but it had not yet made the switch in crucial high-value, high-precision circuit board operations known as surface mounting. As Montreal was coming into effect, the company chanced on a small firm producing a terpene-based solvent that seemed suited to the task, and the two companies jointly developed a process that could be applied to surface mounting. Shortly thereafter, on a site visit organized by the new TOC, a group of experts observed yet another solution: using a machine to solder in a controlled atmosphere could prevent oxidation of the solder joints, eliminating the need to clean the circuit board with solvents in the first place. AT&T, Nortel, and Ford each bought several of the balky machines and collaborated to make them robust enough for their use.[39]

The proliferation of substitutes benign to the ozone layer as well as processes not requiring any cleaning at all took the incumbent ODS producers by surprise. They had expected to profit from the shift away from ODS by selling their preferred alternative, only to find that users' innovations transformed the market under their feet. By 1990, the United Kingdom's Imperial Chemical Industries—one of only two producers of CFC solvents that also sold the cleaning equipment that used solvents—began to cannibalize its own CFC supply business by shifting its solvent equipment sales to devices that used terpenes.[40] In the jargon of business today, the CFC producers had been "disrupted." Fixated on a sophisticated and, it seemed, self-evidently superior product whose development they had mastered, they ignored potential threats from technologies they had dismissed or never considered, until new competitors dislodged them from the market.[41]

But even within sectors such as solvents where progress was disconcertingly rapid, there were occasionally subsectors in which users' requirements were so exacting that substitution had to proceed cautiously. The TOCs would review the requests and propose an exemption, conditional on effort.[42] As many of these chemicals were phased down to zero, the need for these elastic exemptions often grew. Tightening too fast—beyond the limits of good faith, fully informed efforts—invited political backlash.

Perhaps the best example of tightening without overreaching is metered dose inhalers, a drug delivery device that used ODS chemicals as the propellant. Exemptions for metered dose inhalers—with quantities set for each country where they were sold and were critical to the public welfare—were reviewed annually with technical experts, who then determined whether alternative metered dose inhalers had demonstrated adequate and safe performance for each of the drugs they delivered. Once two to three viable alternatives were proven, the exemptions were quickly removed. Industrialists thus could not afford to slack off in the search for alternatives, lest they be left behind when the market shifted.[43]

In some exceptional cases, user-led innovation was inhibited within an entire sector. In refrigerants, for instance, efficiency required that long-lived cooling equipment be precisely matched to the properties of the heat-transferring gas. The resulting codependence between refrigerant producers and equipment makers meant that the latter could not strike out on their own. Nor did the former have incentives to cooperate with each other, since each hoped to develop a substitute with distinctive properties, enhancing its position in the market. The situation was similar in insulating foams, with the additional complication that even if producers of foam-blowing gases and their customers could converge on an acceptable substitute, any substantial increase in the price of the final product could drive demand away from foams entirely to other insulating materials. Because of these entanglements, a cooperative search of any kind was difficult to organize, and these two sectors were among the slowest to innovate.[44]

Nevertheless, there was a profusion of alternatives in almost every sector by 1990, and two years later it was evident that the use of nearly all the original CFCs and most of the halons could be effectively eliminated. As Ted Parson, the leading chronicler of the ozone regulatory history, states flatly, "There was no technical reason that the burst of innovation that began in 1986–1988 could not have happened several years earlier."[45] The barrier then was not the want of fundamental scientific or technical knowledge but rather the reluctance to do the hard work of development needed to make potential solutions actually useful. What made the difference were penalty defaults and government support of cooperation, and with them a shift, sector by sector, of industry interests in favor of a joint search for solutions.

But as the new problem-solving institutions were consolidating and proving their worth, the Montreal regime encountered two limits, each in part a product of its very successes. First, while production and use declined rapidly in advanced countries, they increased in large developing countries such as China, India, and Brazil—none an original party to the Montreal Protocol. Second, as substitutes were found for the original CFCs and halons, it was necessary to expand controls to include new substances.[46] Some of the additions were familiar products, long in use, that simply advanced in priority as the Montreal regime first addressed the biggest problems. Others were new substances created as ODS substitutes that had been accepted despite the fact that they added to global warming, but were no longer tolerable once warming itself became an urgent problem. For Montreal to succeed, the geographic scope and substantive sweep of the regime would have to expand.

EXPANDING GEOGRAPHIC REACH

In 1990, when the parties met in London to adopt their first amendments to the Montreal Protocol, the treaty included only a small subset of developing countries, but still covered 99 percent of global ODS production and 90 percent of consumption.[47] China, not yet a party to the treaty, accounted for a small fraction of the unregulated global total. Six years later, China had become the world's ODS leader, producing 34 percent of the global output and consuming 30 percent. In these years, China switched as well from being a net importer to a net exporter of ODS.[48]

Given this trajectory, it was clear that China and other expansive, highly capable developing countries such as India and Brazil would see joining Montreal as a mixture of risk and opportunity. The risk was that in agreeing to controls, they would need to accept costs or other burdens that slowed the growth of their economies. The opportunity was to mesh their standards and production processes with those rapidly being deployed in advanced countries, ensuring exporters access to developed markets and helping domestic industry stay abreast of technological development. The eventual deal balanced these considerations: developing countries accepted controls, and the reporting obligations that went with them, in return for compensation for the "agreed, incremental cost" of abatement and capacity-building technical support.[49] Funding was to be provided by the MLF, comanaged by developed and developing countries.[50] Violations were to be punished by trade sanctions limiting the imports or exports of ODS along with products using them.[51]

The greatest challenge posed by the agreement proved to be helping developing countries build institutions that could effectively foster and supervise complex abatement projects. The solution was an innovative system, along experimentalist lines, of joint planning and an ongoing review of the national infrastructure of ozone abatement. It allowed national governments and the MLF to engage in close monitoring as well as quick revision of not just individual projects but also the comprehensive system of decision-making by which projects were selected, budgets assigned, results reported, and commitments adjusted.

Under the arrangement, developing countries seeking support from the MLF had to propose a country program describing the institutional and policy frameworks implementing the country's commitment to abatement. The "institutional framework" comprised the entities that would choose projects and monitor the phaseout of ODS, starting with a National Ozone Unit as a central coordinating point (augmented in large countries with regional and local counterparts) and system for tracking progress. The "policy framework" described regulations or voluntary certification schemes encouraging firms and other actors to phase out ODS. Reports on implementing the country programs were to be submitted annually. Taken together, these requirements integrated the review of compliance with control requirements along with discussion of the need for technical support or other forms of capacity building into a single, continuing exchange. Advanced country sponsors were thus reassured that projects stayed within budget and actually eliminated ODS-using technologies; developing countries had, in addition to funding, assurance of timely help with problems ranging from decoding the technicalities of applying for financial support to installing new equipment or building reporting systems.[52]

China played an important role in insisting on and negotiating the MLF, and its experience with the agreement has been emblematic. Exports of refrigerators had declined by 60 percent between 1988 and 1991 as advanced country markets shunned the import of equipment using CFCs.[53] There was no mistaking the thrust of development. Still, the Chinese ministries then directing the economy were divided. The State Planning Commission opposed joining the treaty on the grounds that China lacked the financial and technical capacities to switch to ODS substitutes; the Chemical Industry Ministry was concerned that CFC substitutes would have to be imported, damaging China's nascent chemicals industry and directly threatening plants, recently built, to produce CFCs. On the other hand, the Ministry of Light Industry, representing export interests, and the National Environmental Protection Agency (NEPA) favored joining the protocol. In the end, the high-level State Council asked the NEPA to convene a government-wide cost-benefit analysis of the choices, and the outcome was to ratify the treaty—but only on the condition that agreement could be reached on what became the MLF mechanism for funding. With the MLF in place, cooperation with donors and technical experts helped the Ministry of Light Industry, NEPA, and their allies overcome early failures in implementation, strengthening their hand in domestic politics and reinforcing China's commitment to the ozone regime.[54] As capacities increased, the Chinese found projects that paid for themselves, reducing dependence on the MLF.[55] Similar conflicts surrounding ratification played out in countries such as Brazil, Mexico, Thailand, and Turkey, with similar resolutions and early effects of participation in the protocol.[56] But China stands out in its ability to make use of the possibilities for development that the elimination of ODS afforded. Indeed, in three basic features—deliberately assessing the national interest, tapping into multiple forms of support in addition to funding to transform industrial policy, and accepting a reporting regime of mutual transparency—China anticipated the institutional arrangements that would come to define an emergent global trade regime, which we discuss in chapter 7. Those arrangements, by design, help governments and firms govern common problems by realigning national policy strategies and interests, and increasing transparency and accountability.

In some cases, developing countries embraced Montreal with such determination that they joined advanced country counterparts in ensuring the rapid reduction of ODS. For example, participants in a 1994 study tour to Vietnam organized by the Industry Cooperative for Ozone Layer Protection and UNEP discovered (and closed) an important loophole during a site visit to a plant assembling autos from German and Japanese kits: some of the air conditioners being installed were models using CFCs that had been withdrawn from circulation elsewhere and dumped on the Vietnam market. Another site visit turned up refrigeration equipment imported from Japan that likewise was no longer for sale there or in other advanced countries. Concerned that such intentional or inadvertent dumping could be widespread, representatives from firms such as AT&T and governments such as the United States (represented by Stephen Andersen) and Vietnam composed a letter demanding that multinationals pledge to police their own as well as, implicitly, one another's behavior. By the next day, the letter had been signed by the key ministries of the Vietnamese government, and a few months later, Vietnam, together with the Industry Cooperative for Ozone Layer Protection and US EPA, announced that forty multinationals had joined the pledge. The cooperative equilibrium that had come to dominate the domestic response in the United States and other Western countries to ODS reached the crucial developing countries too.[57]

EXPANDING REGULATORY COVERAGE AND GOALS

The complement to the geographic expansion of the protocol regime was the expansion of its jurisdiction to include new chemicals beyond the short list of CFCs and halons in the original 1987 treaty. The most vexed expansion came in 1992 with amendments to include methyl bromide. The difficulties of phasing out this substance laid bare a design limit that strained but did not break the problem-solving machinery of the regime, and pointed to the need, as we discuss in chapter 5, to complement sector-based variants of experimentalism with place-based ones.

Methyl bromide was widely used, especially in developing countries, as a broad-spectrum pesticide and fumigant in agriculture. An important application was sterilizing the soil before planting in it. While there were substitutes for almost all applications, most of these were crop and pest specific, and often place specific as well; many depended on new technologies and costly practices that could in turn only be effectively deployed with technical assistance.

Pesticides based on methyl bromide, for instance, could be replaced by integrated pest management, in which organisms that naturally prey on the pest are introduced into its immediate environment to control it. However, even when the species of the prey is the same, different predators thrive in different environments. For another example, seedlings can be protected from disease and pests with hydroponic techniques. But hydroponic methods for protecting, say, tobacco seedlings in Zimbabwe cannot simply be transferred to other crops and locations; a breakthrough in one place does not therefore automatically clear the way for widespread emulation.[58] Developing such particularized solutions, proving they work under local conditions with affordable and accessible inputs, and teaching farmers to use them is expensive and time-consuming. The burdens of such differentiated transitions are especially great for developing countries, where research capacities and agricultural extension service may have to be built out to take on the new tasks.

China was particularly aware of the need to field test substitutes for methyl bromide, and secure farmers' acceptance through on-field demonstrations and other methods. At the urging of the Ministry of Agriculture and State Bureau of Tobacco Monopoly, the Chinese government delayed ratification of amendments establishing and tightening controls on methyl bromide because of these concerns.[59] China's hesitations were widely shared. Developing country exporters of other high-value agricultural products such as cut flowers and fresh vegetables resisted controls as well.[60] Producers of methyl bromide in advanced countries, which did not have much to fear from public pressure at the time as activists focused on other industries, further mobilized this opposition in hopes of weakening or averting regulation. Together these actors withheld information from the TOC, overstated the difficulties of switching, and pushed incessantly for critical use exemptions.[61]

In the end, though, the TOC system worked as it did with every other chemical—albeit with fewer industrial supporters, at a much slower pace, and with greater contention. Critical use exemptions in developed and developing countries continue to the present day—nearly three decades after the use of methyl bromide was first limited—precisely because substitutes are so often context specific. A patchwork of institutions encourages local searches; the US EPA and Department of Agriculture, for instance, maintain a website together that gathers current information on "laboratory, field, and on-farm research and technology transfer topics" related to methyl bromide alternatives.[62] Meanwhile, the MLF funds projects—hundreds of them—that demonstrate the feasibility of alternatives, train the trainers who will help implement them, and then orchestrate the actual phaseout.[63]

The persistent difficulties in phasing out methyl bromide reveal a design limit in the cooperative problem-solving institutions of the Montreal regime. The sectoral TOCs work well when the goal is to substitute one standard solution with another—say, to replace an ozone-depleting solvent with an ozone-safe solvent or a production process that doesn't require solvents at all. Whether the search for such general solutions is conducted jointly (through site visits or collaborative research) or in parallel (as firms and other participants investigate possibilities in their respective domains), the findings are pooled and reviewed in the TOC. This open-ended process regularly finds substitutes that the incumbent producers had considered scarcely or never—and superior to the ones they did.

But this system falters when alternatives are heavily context specific. In this case, the limited transferability of results means that ODS users have little incentive to cooperate with each other in searching for solutions. This obstacle, in turn, makes collusion with producers more appealing as a bulwark against change. In these circumstances, determining the solution through centrally coordinated search and review can be, at best, the frame as well as starting point for local investigation of what exactly works in a given place; the crucial, concrete problem-solving takes place on the ground. A sectoral TOC can take note of these efforts, but the local context is too important to allow many useful general rules for replacement of the offending chemicals and practices. In chapter 5, we will see that many of the most pressing current problems of decarbonization also require place-based solutions.

Climate Change: Wrong Turn to Kyoto

In contrast to the overarching experimentalist successes of Montreal, cooperation on climate change evolved along a different track—one that created institutions that were rigid and tightly controlled by diplomatic processes, and thus did not allow for much experimentation. In short, despite extensive diplomacy and growing attention to the problem of global warming, they didn't do much problem-solving.

Climate came on the global agenda in the late 1980s as a natural progression from the concern with ozone. At the time, many climate diplomats spoke of the Montreal Protocol as the model for a treaty on greenhouse gases. They regularly invoked the agreement and its early successes as a demonstration of how the world could respond with dispatch to a planetary threat, although they did not call much attention to how the Montreal Protocol pointed the way to new institutional forms of cooperation.[64] Where uncertainty was recognized, it was nearly always uncertainty in the science of warming rather than uncertainty regarding the way forward. And this limited recognition of uncertainty was usually coupled with calls for more research as opposed to calls for institutional forms to better manage the uncertainty.

Despite the perception that there was (and should be) continuity in the forms of international cooperation used to tackle these two problems—ozone and warming—the situation had changed fundamentally since the early days of Montreal. The background conditions that had induced collaborative problem-solving in the case of ozone were largely absent in later climate diplomacy. While penalty defaults had accumulated in the fifteen years it took to arrive at the ozone regime, nothing similar had happened with respect to climate. Efforts to establish a credible, global regime for regulating illegal logging and other threats to forests (a major source of global emissions) were a centerpiece of the 1992 Earth Summit in Rio de Janeiro, for instance, but those efforts failed—a sign to firms and governments on the front lines that they might not have to worry much about the consequences of their behavior. In the case of ozone, the high likelihood of regulation or reputational consequences for intransigence made inaction or active resistance a risky strategy. By contrast, there were few such expectations or incentives in the formation of the regime to slow global warming. Starting in the 1990s, European governments and activists at the forefront of the movement for a climate regime focused increasingly on the need for action; in time, they developed a few initiatives on their own to target climate-related activities such as bans as well as controls on imported tropical lumber and palm oil. But outside a handful of industries, few firms felt much need to change. In most industries, if firms discussed climate change at all, it was usually as a matter of corporate social responsibility and public relations, not a redirection of corporate strategy. (Today, climate penalty defaults are a lot bigger in a few places in the world—but only after more than three decades of policy efforts.)

Meanwhile, convening power over the climate negotiations—along with de facto veto power over their outcome—was passing to the developing countries, which were deeply skeptical of cooperation with multinational firms and the advanced countries in which they were headquartered.[65] UNEP had played the central organizational role in the ozone negotiations; the activism of its largely scientific and technical staff appealed to counterparts the world over. Governments inclined to the same kind of activism on warming—member states of the European Union such as Germany, the Netherlands, and the United Kingdom—sought to have UNEP organize the climate change regime as well. On the other hand, as domestic political support for action on environmental issues became more polarized, the United States had turned more cautious. It lobbied to have negotiations framed by the Intergovernmental Panel on Climate Change (IPCC), which had been formed in 1988 to assess climate science and policy, in the hopes that talks would narrowly focus on scientific and technical issues—above all, on the economics of climate change.[66]

Developing countries, fearful that rich industrialized countries would use climate change to hold back their economic growth while disclaiming their own historical responsibilities for a warming world, rejected these alternatives.[67] Instead, through a series of formal decisions, they placed authority for launching the negotiations in the United Nations General Assembly, where they could confidently exert control themselves.[68] The developing countries had, in the 1960s, formed the G77 caucus to represent their interests in the UN General Assembly and outside it.[69] The caucus was large and powerful, but it encompassed an unruly array of interests—from those of economic powers such as China with expansive industrial economies and export interests, to those of oil exporters that feared regulation of their product, to the world's least developed countries, including low-lying island nations with little to export and literally everything to lose as the sea level rises with global warming.[70] On one front, however, the developing countries were united: their determination that they themselves be exempted from emission controls and the full burden of responding to climate change be assigned to the advanced countries. In any case, since the UN General Assembly, by convention, makes all decisions by consensus, the climate change regime that would emerge did the same, thus giving each developing country a veto over any proposal, no matter its origin. The advanced countries had no choice but to go along.

It took only some fifteen months from the formal start of negotiations early in 1991 to produce a nearly complete text of a climate treaty: the UNFCCC.[71]

The document was able to be drafted so quickly in large measure because it reflected only what was agreeable, with everyone knowing that consensus yielded the lowest common denominator. The developed country parties agreed formally only to "take the lead in combating climate change and the adverse effects thereof."[72] But aside from some minimal reporting requirements, the UNFCCC did clearly protect developing countries from meaningful—costly—obligations. "Developed" and "developing" categories were defined, and every country was categorized accordingly. The eventual commitments to cut emissions, whatever they turned out to be, would apply exclusively to the developed countries.

Diplomacy on climate change evolved quickly from this first framework of the UNFCCC. For more than a decade, those diplomatic energies focused on creating and managing the Kyoto Protocol. As evidence mounted that the Kyoto system was faltering, it proved difficult to gain consensus on alternative courses. So the diplomats soldiered on, only finally to pause and regroup when all the side deals and subagreements needed to hold together a consensus proved too fragile. That moment of pausing came, as we will see, in 2009, at the annual Conference of the Parties (COP) in Copenhagen. The formal agenda at Copenhagen was to reach agreement on a replacement for the Kyoto Protocol. What resulted instead was an awareness that no agreement was universally agreeable. Copenhagen ended in disarray—supplying an opportunity for fresh thinking.

BERLIN, KYOTO, AND BINDING TARGETS

Because it was largely devoid of content and inoffensive to practically any country, the UNFCCC was quickly and widely ratified. Within two years it entered into force, and the first formal COP was convened in April 1995 in Berlin. Three decisions were taken in Berlin that with little or no deliberate consideration of their broader implications, conclusively ruled out experimentalist problem-solving on the Montreal model, and entrenched the implicit commitment to fixed targets along with the explicit, rigid division of obligation between developed and developing countries.

First, in a declaration known as the Berlin Mandate, the conference proclaimed the indeterminate and hortatory goals of the UNFCCC inadequate, and pledged to replace them with unambiguous and ambitious commitments to targets as well as timetables for their achievement by the third COP, scheduled for 1997 in Kyoto.[73] The developing countries insisted that the more precise restatement of purpose "not introduce any new commitments" for them.[74]

Second, the Berlin meeting restricted membership in UNFCCC committees and all subsidiary bodies to "government representatives."[75] This maximized the control of the state parties over the evolution of the agreement—reducing the risk that independent experts would gain inconspicuous and unchecked influence, but at the price of foreclosing the participation of academics and leading industry experts that made concrete problem-solving possible under the Montreal Protocol.

Third, Berlin adopted a funding program offering basic assistance to developing countries. But it included no support for the kind of comprehensive institution building that was emerging with the Montreal MLF. In the case of ozone, that support was intended to help developing countries not only comply with treaty obligations but also take advantage of the possibilities for technology transfer along with the coordination of standards that facilitate economic growth and exports, and quickly blurred the lines between "developed" and "developing." Instead, under pressure from donors, the UNFCCC turned to a new mechanism, the Global Environment Facility, recently created as a kind of all-purpose vehicle for funding international projects. The Global Environment Facility was small in relation to the scale of the climate problem, and concentrated, as many international aid organizations do, on the distribution of the funds allocated to it and ensuring that the monies are dispersed on schedule for the agreed-on purposes.

In yet another sign of wariness about any agreement with impact, Berlin was remarkable for *not* deciding something that the first conferences of treaty parties routinely do decide: rules of procedure. Because of disagreements over alternate voting rules, the COP never agreed on voting systems, and by default—again following UN General Assembly convention—all decisions were (and still are) taken by consensus. This meant that the ambiguities, introduced to secure agreement, accumulated; discrete discussions to resolve the ambiguities proliferated; conference decisions, obliged to address (or at least not unsettle) the partial clarifications became package deals; and the package deals became so unwieldy that nothing of substance could be changed. The practice of consensus decision-making continues to the present.

Why did developed countries and the NGOs that were most motivated to address climate change go along with this framework? To some degree they had little choice because alternative approaches had proved unacceptable to the consensus needed for any diplomatic action. Moreover, the strict system of targets and timetables required by the 1995 Berlin Mandate, which became the centerpiece of the Kyoto Protocol, aligned with the core policy goals of most developed countries. For Europeans and NGO activists—the most reliable advocates for cutting warming pollution—targets and timetables were thought to be the only way to guarantee pollution would go down. The importance of binding country-specific limits was, they thought, one of the lessons from the Montreal experience.[76] At the time, few were focused on the kinds of experimentalist institutions that made it possible for Montreal to impose such strict standards while keeping those standards in line with what was feasible.

As many in the developed countries saw it, especially in the United States, Kyoto's inclusion of a rigid system of fixed pollution reductions could become the foundation for a new kind of international pollution control market. The turn to market mechanisms in the United States arose from broad criticism of the bureaucratic, New Deal state, and in environmental regulation, dissatisfaction with the Clean Air and Clean Water Acts of the early 1970s.[77] For critics of the climate regime, the alternative was to shift the burden of finding an effective solution from government regulators to firms. These ideas were being put into practice in the 1980s through the creation of various kinds of "bubbles" (trading regimes allowing the transfer of permits between installations of a single facility, for example, or between old and new facilities) and a nationwide market for sulphur pollution (the leading cause of acid rain). In practice these markets had mixed results at best, as we will see in chapter 4. But politically—and for economists, conceptually—the ideas proved almost irresistible.[78]

Climate diplomats from developed and developing countries quickly and successfully introduced these promarket ideas into the Kyoto discussions. In principle, emissions trading is even better suited to controlling greenhouse gases than controlling pollutants with regional or local effects such as sulphur dioxide. Given huge global variation in the marginal costs of greenhouse gas abatement, the potential gains from trade were also expected to be huge. Moreover, carbon dioxide pollution is not directly harmful to humans, meaning that a standard concern regarding pollution trading—the rise of geographic "hot spots" of noxious pollution—did not apply.[79] For once it seemed that the attributes of the pollutant in question allowed theory to be applied to practice.

Largely at the urging of the United States, supported by Brazil, Japan, and a growing number of other countries, the Kyoto Protocol provided for no less than three emissions trading programs. "Joint implementation" allows for trade between developed countries that had adopted emission control limits, meaning that countries in central and eastern Europe—formally classed as "developed," but with relatively low abatement costs because their economies had collapsed with the end of the Soviet Union—could sell their surpluses to advanced country counterparts where abatement is expensive. A second mechanism allowed developed countries to pool their commitments if they were part of a broader economic and political union—a feature that allowed EU nations to treat themselves as one and create an EU-wide emission trading system (ETS). Third was the Clean Development Mechanism (CDM), which allowed investors from developed countries to finance emissions reductions projects in developing counties in return for tradable credits. The CDM was seen as the dress rehearsal for a truly global emissions trading system.

In short order, emissions trading also became the centerpiece of the Clinton administration's argument for entering the Kyoto agreement. Early in 1998, a few months after negotiations concluded, Janet Yellen, then President Bill Clinton's chair of the US Council of Economic Advisers, asserted that while "a global solution" was "critical to the global problem of climate change," it was equally vital that global efforts be as efficient as possible. Because the United States had prioritized this goal in negotiations, Yellen maintained, the treaty text gave a cost-conscious public and private decision makers flexibility in when targets were met, what kinds of emissions had to be reduced, and where on the globe the reductions were to take place. Deadline flexibility allowed the parties to reach reduction targets over the course of a five-year "budget," giving them control over the exact timing of their programs. In terms of the impact on warming, the exact annual flow of emissions didn't matter much, on this line of thought, because the total warming was a function mainly of the accumulation of pollution over time.

Scope flexibility, meanwhile, left it up to the parties to choose which greenhouse gases to limit by including all six in the agreement. The Kyoto Protocol included exchange rates that allowed the parties to convert different greenhouse gases into common units. Reducing one kilogram of sulfur hexafluoride, a pollutant with superwarming characteristics, would count for as much as eliminating 23,900 kilograms of carbon dioxide. The exchange rate reflected these differences in potency, and would let governments and firms, on their own, decide where it was cheapest to concentrate their efforts. This scope flexibility was further increased by counting the creation of carbon sinks through afforestation and reforestation as equivalent to pollution reduction, although the treaty put some limits on that flexibility.

Finally there was geographic flexibility—arguably the "most important of all" according to Yellen (and most economically minded analysts at the time)—with clean development projects holding special promise of being "quite cheap" measured by the cost per ton of emissions avoided. At the time, developing countries already accounted for half of global emissions—a share that was rising—and geographic flexibility was intended to create incentives to control these sources of pollution lest they keep growing unchecked.

Support for this market-oriented approach was seemingly bipartisan. Officials from the administration of George H. W. Bush, which before Clinton had negotiated the UNFCCC, celebrated the market-oriented approach of Kyoto; if anything, they were critical that Kyoto didn't go far enough in making full use of markets.[80] Yellen noted that these views had "long been championed by economists interested in increasing the efficiency of [climate] protection." Some twenty-five hundred economists from academia, industry, and government had put forth a similar position in a letter calling for action on climate change the year before.[81] It would have been discordant, amid the celebration, to observe that the very flexibility that made Kyoto efficient on paper removed the constraints to solve particular pollution problems—involving certain gases, in given industries, in various places, and under deadlines. It weakened the incentives to search for solutions beyond the lowest-hanging fruit.

Putting this all together, by about the year 2000, the Kyoto Protocol had reached a roughly similar stage in legal evolution as the Montreal Protocol had reached in 1988. A treaty was in place, and governments were beginning the process of ratification. The question of implementation now loomed. It is here where the divergences between ozone and the climate became most apparent, for the Kyoto treaty lacked the experimentalist mechanisms that would be needed to engage firms and governments as they tried to put commitments into practice. It lacked, in particular, the mechanisms needed to obtain, organize, and feed insights from implementation efforts—successes and failures—into the setting as well as adjustment of commitments. Instead, the architecture of Kyoto was based on the assumption that implementation would occur in the traditional sense: standards had been set in the treaty, and it was then a matter of governments and firms on their own to align their behavior and comply.

The setting of emission targets and timetables was designed to simplify implementation. Allowing for the trading of emission credits was intended to make implementation even easier by shifting the burden for decision-making away from bureaucrats and to the actors directly involved in projects; the market would settle questions of which activities would be controlled and where investments would flow. (For some countries, trading was merely a windfall to be maximized. Russia, in particular, sat on huge volumes of extra credits because the country's emissions were far below its Kyoto limits thanks to post-Soviet economic collapse.[82] Implementation for Russia involved no problem-solving, but just waiting for a surplus buyer to appear.)

The most revealing problems of implementation came with the CDM trading scheme involving developing countries. By design, the CDM assumed there would be no countrywide experimentation and no countrywide shift to the best-available technologies, such as occurred in China, Vietnam, and many other countries under the Montreal Protocol. Figuring out the volume of credits to be awarded required a calculation that seemed simple to implement, but in practice revealed why implementation in the traditional sense was unworkable. In particular, awarding CDM credits required figuring out the level of emissions from the country that would have occurred with and without a project. Credits were to be awarded in proportion to the difference. But since these two worlds—with and without a project—could never be observed simultaneously, in fact the CDM created incentives to hide information, and avoid economy-wide policies or shifts to the best-available technology. Only by inflating emissions in the counterfactual world was it possible to maximize the flow of credits. Both investors and host countries shared this interest in overestimating the baseline as well as exaggerating the reduction gains from each project.[83] The market was soon flooded with credits of dubious integrity. Worse, this approach created perverse incentives for host countries to create the appearance of high emissions only to maximize credits when emissions were lower. Absent the institutions of experimentalism, which could scrutinize every project while shifting whole economies to lower emission trajectories, it was administratively impossible to distinguish real problem-solving from accounting tricks. This challenge of administering a system of complex project plans, overseen by incomplete third-party monitoring, and steeped in information asymmetries and perverse incentives, was alarmingly familiar to administrative lawyers.[84]

Countries responded to the CDM problem in many different ways, and as those responses accumulated, the unworkability of the Kyoto approach became apparent.

Japan turned to the CDM as a last resort, but embracing it came with disastrous industrial and political consequences. The country, which had hosted the Kyoto event, was making good faith efforts, sector by sector, to cut emissions—a challenging task because most of Japanese industry was already at the global technological frontier and thus very clean. Only when it was clear that these would not suffice to meet treaty obligations did the country resort to buying credits, mostly from China. In short order Japan realized it was subsidizing competitors, and the country's industry soon soured on Kyoto. Japan remained a formal member of the treaty—leaving would be too costly diplomatically—but as a practical matter, Kyoto didn't induce as much implementation as it did backlash.

In the European Union, too, the CDM's conceptual appeal proved unworkable in practice. In tandem with Kyoto, the European Union had launched the ETS, which quickly became the world's largest pollution market. For most of its history, the ETS failed to have much impact because prices were low—thanks in part to Kyoto-friendly rules that allowed EU firms to buy CDM credits to offset their obligations at home. As these cheap credits flooded in, the European Union soon learned of the huge variation in the quality of CDM projects. Fixing that problem would require the ability to inspect every project individually—a task that proved to be essentially identical (if not more onerous) than an administrative command-and-control system. So the European Union simply phased out the use of CDM credits; it took more direct control over the supply of permits into the ETS in order to drive prices higher.[85] Years later, Europe did build a program for emission controls with many experimentalist elements; doing that, however, required insulating the European Union from Kyoto.

In the United States, the CDM had been a major selling point for Kyoto: it was the embodiment of the systems that Yellen imagined in her testimony. As it turned out, even with the purchase of cheap credits (partly from the CDM and mainly from even lower-cost suppliers, the Russians, it was hoped), there was no practical way the United States could comply.[86] But that worry was soon moot. In 2001, the treaty was caught up in the new Bush administration's growing skepticism of international institutions, especially those that might constrain US energy policy.[87]

BEYOND KYOTO

Evidence mounted that Kyoto was not working.

Emissions from developing countries were rising faster than expected. The CDM had little impact on those emissions, and by some assessments it actually raised baselines and thus increased emissions. The use of targets and timetables, wrongly thought to be the lesson from Montreal, was being questioned—precisely because meaningful targets and timetables were unknowable ex ante, and thus implementation in the traditional sense could not occur.[88] The differences between implementation in the traditional sense and real problem-solving were becoming apparent. But it is one thing to raise concerns, and quite another to find institutional solutions under conditions of consensus.

Formally, the opportunity for a fresh approach was offered by the fact that the Kyoto emission targets were scheduled to run over the years 2008 to 2012. Hence a new treaty would need to be in place well before 2012, and a new diplomatic process was launched with the goal of having that treaty adopted by late 2009 at the annual COP meeting hosted that year in Copenhagen.

Awareness of troubles had led, in the run-up to Copenhagen, to many new ideas that might improve problem-solving. Yet the ideas were inchoate and often speculative—far from commanding the consensus needed to chart a decisive new course. On the problem of engaging developing countries, it was widely agreed that more funding would be needed; at Copenhagen, most deliberation focused on a pledge for $100 billion per year of new climate finance. But agreement splintered when it came to how those funds would be counted, let alone spent. New ideas for framing emission control pledges—by developed and some developing countries alike—were also in the air, but no plan had emerged for codifying or checking into those pledges, let alone their legal status. And on many other fronts, the same story emerged—agreement that the existing approaches were not working, but not on a plan for reform.

In the end, Copenhagen proved an anticlimax. The goal was to set a new framework for cooperation that included bigger commitments for developing countries along with bigger funds to help them pay the cost. The visions were partial and conflicting, though, and the package deals harder to put together. Diplomats got close to reaching a deal in Copenhagen, but their ambition was limited to securing a deal on what was agreeable rather than achieving a vision for Montreal-style problem-solving that could be implemented in the case of climate change. Ironically, it was a small group of the least developed countries—upset at not being "in the room" when the major deals were done in Copenhagen, yet largely irrelevant to the global emissions picture—that formally blocked the consensus needed for Copenhagen. But for their pique, climate diplomats might have succeeded again in maintaining what was seen, increasingly, as the illusion of progress.[89]

Copenhagen was nevertheless cathartic. The hard line between developed and developing countries, politically expedient in the 1990s, had become untenable. By then, developing countries were responsible for essentially all the growth in global emissions; for most, development strategies had become more reliant on energy-intensive heavy industry.[90]

The shock of Copenhagen opened the door for other options to be explored—not so much as a matter of strategy, but by default. Only after Copenhagen did it become widely accepted that so much effort had been invested in just one idea, and that idea didn't work.

The Opportunity of Crisis

Analytically, our recounting of the histories of ozone and the climate turns on two comparisons.

The first comparison is of the behavior of firms producing and using ODS before and after they were subjected to the experimentalist regime of the Montreal Protocol. This comparison lacks the rigor of a natural experiment, but it is highly suggestive. The abrupt improvement in problem-solving capacity follows directly on the introduction of the experimentalist elements of the Montreal regime—the incentives that disrupted the status quo, and required a search for new solutions or the adoption of improvements along with methods for evaluating alternatives and updating goals. Searches that ended in an impasse before Montreal was in place, or never began, produced striking results once the regime was in operation. These results were evident not just in advancing the frontier of ozone-benign technologies but also in ensuring rapid widespread adoption as whole industries and countries shifted to the best-available technologies.

The second comparison is between the effectiveness of direct problem-solving in Montreal and the ineffectiveness of the climate regime. Comparisons across regimes, where so many factors are in flux, are necessarily blunt and speculative, of course. But like the before-and-after comparison with the creation of Montreal, the comparison of regimes points to the effectiveness of experimentalist mechanisms. The architects of the climate regime separated goal setting from implementation and tried to put the latter under the direction of markets. The assumption was that this would make each part of the process easier. In fact it made both harder. As the goals were to be binding and fixed, they had to be precise, but precision was elusive and contentious. By the same token, execution by markets presumed the availability of information that proved hard to come by. Trickery flourished in its absence. By contrast, the experimental approach of setting general goals to encourage a broad search, and using the search itself to generate a shared understanding of what was becoming possible, made it easier to reach initial agreement and consensually adjust commitments while executing them.

To crystallize the lessons of Montreal's success—and the failures of Kyoto—we need a firmer grasp of why experimentalism works and the incentives that encourage it. To that end, we turn in the next chapter to the theory of experimentalism as a distinct kind of decision-making under uncertainty.

3

Theory of Experimentalist Governance

In 1977, business historian Alfred D. Chandler Jr. published *The Visible Hand: The Managerial Revolution in American Business.*[1] The book tells the story of how successive generations of managers learned to make mass production into an engine of economic growth and corporate expansion. In Chandler's time, the mass production firm had an air of inevitability, as though it were history's final word on the question of economic efficiency. Success was taken to be self-perpetuating. A firm at the forefront of its industry—whether in cars, chemicals, or breakfast cereals—used the most efficient methods of production to earn enough to invest in the next round of expansion, and leveraged dedicated research laboratories and marketing staffs to mold the future in its image. Incumbency was a prize, and winners took all.

Twenty years later, this confidence had begun to wane. In 1997, Chandler's Harvard Business School colleague Clayton M. Christensen published the *Innovator's Dilemma: When New Technologies Cause Great Firms to Fail.*[2] Rather than boldly setting the terms of the future, the great corporation was now portrayed as perpetually vulnerable to disruption by the development of new technologies and markets. In this new world order, incumbency was a trap. Attempts at foresight—whether through sophisticated tests of the market or tried-and-true customer surveys—only produce a false sense of security. There is always some marginal technology, ignored by experts or dismissed as irrelevant, that can pose a challenge to dominant solutions. Instead of trying to anticipate developments or head off disruptions, firms must acknowledge deep uncertainty and open themselves to surprises.

These two books neatly encapsulate an epochal shift in governance still underway today across large economic and political institutions in the most dynamic parts of the world economy. For much of the twentieth century, firms and regulatory bodies operated under the presumption of a stable environment—or at least with confidence in their capacity to maintain stability. In the mass production economy, the trajectory of development was always more or less clear. With the stark rise of uncertainty in recent decades, however, both large corporations and government agencies in advanced sectors have mostly abandoned the pretense that they can predict the future. Instead they aim to make the most of conditions they cannot foresee, inviting a clash of multiple design approaches and embracing flexible organizational structures that can be reorganized as circumstances change. In a word, experimentation is now the order of the day.[3] Since the turn of the millennium, many economic and political institutions have developed robust new methods to learn from the surprises as well as shortcomings that arise when production lines and regulatory frameworks—now built as much to reveal stress as to shield against it—meet unanticipated conditions.

The threat of climate change and challenge of decarbonization are surely among the most testing circumstances these institutions face. The central argument of this book is that institutions at all levels will almost certainly fail to meet this challenge if they do not rapidly develop these capacities for experimentation and learning. We must, in short, inaugurate a new era of experimentalist governance. But to do so, we must get clear about exactly what elements of emergent institutions contribute to their experimentalist success. This chapter distills the principles and practices of these new forms of organized provisionality, all in the hope of aiding their adoption where they are so acutely needed.[4]

Experimentalism is at once an organizational structure that routinizes self-monitoring and mutual correction between ground-level problem solvers and those orchestrating their efforts; a form of deliberation that uses doubt and disagreement to progress despite uncertainty; and a set of incentives that, by making it risky to bet on the status quo, encourages actors to adopt such structures and deliberative routines when the public interest requires. In this chapter, we examine experimentalism from the perspectives of each of its three aspects or components—governing organizations, deliberation, and incentives. But first we give some preliminaries, situating experimentalism within larger debates about governance over the last few decades.

Governance

If democracy is government by and for the people, governance is the mesh of public and private authorities—formal laws, informal procedures, and shared conventions—that determine in practice who regularly participates in democratic self-determination and under what conditions. To give only a few examples, it is governance that determines whether a government provides transportation or health services itself, or else contracts with private providers. It is governance that determines whether regulation is adversarial or cooperative, and what those terms actually mean. And it is governance that determines the terms and institutions through which workers can collectively bargain.

Because governance links formal law to informal practice and public to private decision-making, it is sensitive to changes in economic and political conditions often well before such changes prompt reconsideration of legislation or regulatory rules—if they ever do. In short, law tends to lag where governance leads. For this reason, governance is frequently the language in which think tankers, academics, and business leaders broach proposals for reform too urgent to be postponed, yet too speculative to be codified into law. From this point of view, governance is a kind of test bed for urgent and often highly consequential adjustments to new circumstances.

The rise of uncertainty is recasting the nature of the governance problem. For most of the last century, governance debates have been concerned with determining the identities and aligning the interests of principals and their agents. On this model, principals—whether the sovereign people, a legislative body, government administrator, controlling owner of a corporation, or corporate manager—have plans and projects. But to realize these plans, principals must rely on agents with interests of their own. If the background conditions are fixed, the principal's problem is to devise an incentive system that induces agents to spend their efforts in achieving their goal rather than using the discretion their position affords to disguise self-interested behavior as dedication to the project. For instance, tying manager compensation to stock price increases links the interests of principals and agents when the principals are shareholders.[5]

When the background conditions are changing, however—as is typical of the episodes that provoke a concern with governance—the problem of aligning interests becomes entwined with the larger problem of determining which actors should be principals in the first place. In extreme cases, the roles can be reversed. For the architects of the New Deal, for instance, shareholders were too dispersed and removed from day-to-day operations to exercise control of the corporation. High-level managers became the principals of the corporation. The New Deal checked their discretion by subjecting them to regulatory oversight and enacting a reform of collective bargaining that allowed trade unions to exercise what later came to be called "countervailing" power. But decades later, as managers built empires by expanding into lines of business for which they were unprepared, activist shareholders, claiming to be the true corporate principals, broke conglomerates into separate pieces, paying for the takeovers with massive debts.[6] Today shareholder supremacy is itself under attack for shortsightedly and unjustly favoring equity over other stakeholder interests, such as those of employees or suppliers.[7] In the world of mass production, solutions changed with the identities of who is accountable to whom, but all addressed the principal-agent problem.

Even as these debates continue, the principal-agent relationship has been breaking down under uncertainty—first in pockets of the economy (like automobiles and semiconductors) and areas of regulation (like pharmaceuticals or air pollution) especially exposed to rapid change, and then in fits and starts more generally. In a risky world, actors can assign probabilities to outcomes and incentivize behavior that leads to the desired ones. Under uncertainty, it's impossible to anticipate what the outcomes will be and hence impossible to assign them probabilities.[8] In these circumstances, the challenge for governance changes radically. Under uncertainty, no actor alone can formulate plans with the precision and confidence necessary to engage agents for precise tasks, let alone devise methods to incentivize and hold them accountable.[9] Conception cannot usefully be separated from execution. Actors instead have to collaborate, defining projects in the very process of trying to carry them forward, and using such progress as they make to reassess the feasibility of the undertaking along with the capabilities and reliability of their partners. In a stable world, in other words, agents execute steps in the plans of their principals, and the fact of their interdependence is covered over by the serviceable fiction that the principal is in control. Under uncertainty, however, planning and execution are inextricably connected; pooling their knowledge and experience, actors use the execution of provisional plans to revise their joint goals. Their mutual dependence is as open as the fallibility of their projects.

Over the last few decades, in advanced sectors of the economy and public administration, this kind of collaboration has been institutionalized and given legal form, keeping cooperating actors accountable to each other despite the fluidity of their relations and transience of their plans, and thus protecting them against the vulnerabilities that their interdependence creates. Organizations are designed so that anomalies and surprises touch off an investigation of possible improvements rather than efforts to enforce the existing structure against newly identified risks. Regulation and contracts, long intended to be proof against every imaginable contingency, are likewise being reconceived as open to learning through use. The form of administrative decision-making is shifting as well, from the promulgation of rules to the issuance of guidance, in recognition of the impossibility of certitude and therefore the acceptance that directives will routinely need to be revised. In the next section we sketch these developments, which together create the resources for the rapid learning and self-correction required for the vast transformation under uncertainty necessitated by decarbonization.

Experimentalism as Organizational Structure

The best way to understand experimentalism as an organizational principle is to see how it emerged as a response to the transformation of the mass production economy. In this section we sketch, in necessarily bold strokes, how we went from a world of self-perpetuating efficiency gains along a familiar trajectory to a world in which an open acknowledgment of uncertainty as the sheer inability to anticipate future states of the world is the precondition for success. In such a world, resilience is achieved not by buffering organizations against known risks but rather by designing them to be easily reconfigured when circumstances unforeseeably change. Resilience comes not by making organizations as self-sufficient as possible but instead by equipping them to cooperate with outsiders as needed. This transformation, pioneered and today still most conspicuous in the way the production of goods and services is organized and products are designed, thus goes hand in hand with more collaborative forms of contract and regulation as well.

THE FATE OF THE MASS PRODUCTION FIRM

The efficiency gains of the mass production corporation came through economies of scale: the greater the volume of output, the lower the cost of each unit produced.[10] These economies of scale were achieved by successively larger investments in ever more specialized machinery and organization, but they came at the price of increasing vulnerability to shocks. The ever-larger investments were only profitable if the market for their increasing output grew apace; otherwise, extremely expensive equipment sat idle. As behavioral economist Adam Smith observed long ago, in achieving economies of scale, the division of labor is limited by the extent of the market.[11] Disruptions in supply could also be as costly as shortfalls in demand. If the supplier of a key part suffered an accident or used the buyer's dependence on a highly specialized product to extract better contract terms, production in a whole plant could be idled.

Chandler saw clearly that the mass production firm was organized and managed to reduce these risks. He did not foresee that in the long term, the precautions made it even more fragile and susceptible to shocks.

To minimize the risk of idle capacity, the firm looked past business cycle fluctuations and carefully calibrated expansion to the expected growth in the long-term demand for its products. Increases in the minimal, efficient scale of plants could thus be matched to increases in the reliable market. To minimize the risk of disruption in supply, the firm vertically integrated, making all of its key components and systems itself, and contracting for the rest with outside suppliers to produce parts to its exact specifications. To minimize the risk of disruption in the day-to-day flow of production, the firm maintained large inventories of work-in-progress on the shop floor. If a machine broke, the part it should have been making could be supplied from this stock until it could be fixed. These choices in turn entailed extremely long planning cycles. Exhaustive upstream planning was assumed to be the only way to avoid errors that would be discovered, if at all, at great cost during the downstream implementation of designs.

In recent decades, this strategy—and the precautions to which it gave rise—was first undone and then reversed in the advanced sectors of the economy and government. Starting in the 1970s, economic shocks—price spikes and quantity disruptions in essential inputs like oil and grains, along with the emergence of new competitors from developing countries—undermined confidence in the long-term planning on which steady expansion had depended. With the information technology revolution, the pace of innovation only increased, and as breakthroughs in one domain allowed advances in others, the trajectory of technological development within particular industries also became unpredictable. In-house suppliers, long at the forefront of their fields, were suddenly found to be using outmoded production technologies or making outmoded products.

LEARNING FROM CONTROLLED FRAGILITY

Amid the turmoil and confusion, it was gradually discovered that trying to minimize risk by exhaustive planning and buffering production against disturbances—effectively hiding alternatives and problems from view—could be extremely wasteful. Against the prevailing intuition and the experience of a century, it turned out that opening organizations to alternative designs early on and making the setup of production vigilant to stress, so that even small disturbances are immediately registered and the vulnerabilities they may presage quickly identified, could yield both better products and productivity gains superior to those achieved by increasing specialization, all without the penalty of increasing rigidity that specialization entailed. A kind of deliberately induced and carefully controlled institutional fragility, in which strain is immediately apparent, and its appearance triggers the investigation of causes and countermeasures well before it ramifies into catastrophe, showed the way to resilience.

Vertical disintegration was a key step. Firms sold their captive suppliers. General Motors, for example, long an exemplar of the logic and success of vertical integration, produced 70 percent of its parts in the 1980s and purchased the rest on the market. Twenty years later, the ratio was reversed.[12] But relations with independent suppliers changed radically as well. Instead of forcing suppliers to compete to make a fully specified part at the cheapest cost, automakers chose to collaborate with outside suppliers at the outset in the design of new products, when their specifications were open-ended. General Motors was again—improbably, given its long record as a bully in the market—a bellwether; it acknowledged that the tables had been turned, and to have access to the most innovative technology, it has to compete to be the customer of the best suppliers, not the other way around.[13]

A notable feature of this collaboration with outside suppliers is that it's carried out in parallel rather than in series—concurrently as opposed to sequentially through time. In this way, choices within each design area prod reflection on the desirable characteristics of other, complementary systems, extending the range of alternative conceptions of the product that can be canvased. It had been assumed that widening the scope of the search in this way would jeopardize the reliability of the final design, but it turned out there was no such trade-off. The comparison of alternatives within and across particular design domains resulted in such penetrating evaluation of their strengths and weaknesses that the final result was a better, more reliable design, arrived at faster than before because the process was concurrent.[14]

The organization and management of the flow of work on the shop floor was likewise transformed. Uncertainty dramatically increased the cost of holding large buffer stocks of work-in-progress inventory as a hedge against disruptions. Firms thus eliminated the buffers, at the limit producing one piece at a time. In this just-in-time or lean production, a breakdown at any one station stops all operations immediately; production can only resume when the disruption has been traced to its source and corrected, reducing the underlying variations and imprecisions that degrade quality. And just as with the expansion of the search for product design, this strategy had surprising implications. It had been assumed that the price of increased quality would be an efficiency-reducing slowdown in production. But it turned out that by building in this requirement to root out vulnerabilities, the firm reduced the incidence of breakdowns to previously unimaginable levels, eliminating the underlying sources of variability, and increasing quality and efficiency together.[15]

The vertical disintegration and short learning cycles that are characteristic of modern manufacturing are also coming to characterize other sectors of the economy. These developments are most salient in agriculture, especially in the diffusion of precision or no-till planting, which does away with plowing. Seeds are inserted essentially one at time to a depth and with a dosage of fertilizer adjusted to the conditions of each "pixel" of land. This method avoids soil compacting and erosion; the results are monitored pixel by pixel; and the conditions are adjusted after each planting to increase the yields by taking account of microfield variations in drainage or soil. Precision agriculture, like lean production, generates pressure for continuous improvement, and as part of that, purposeful differentiation of inputs, opening the way to the introduction of green technologies such as sprayers targeted to the low-dose application of pesticides and herbicides.[16] We will return to these developments in case studies of the reform of dairy farming in Ireland in chapter 5.

TOWARD COLLABORATIVE REGULATION

These changes in production methods have profound effects on regulation as well.

Even in a stable world—one in which regulated entities in particular domains are relatively homogeneous—it is difficult to write enduring, generally applicable rules. The difficulty is that private actors have a clear understanding of the potential effects of regulatory measures on profits and the choice of technology, but regulators do not. Incumbent firms exploit this information asymmetry to escape costly requirements, and at worst, manipulate regulations to enhance their returns and bar new entrants. As the principal in this situation, the regulator seeks to elicit from agent firms the information needed to protect the public, without being "captured" or ceding control to the better-informed, regulated party.

Uncertainty does not eliminate the information asymmetry between the regulator and regulated entity; there are still important things a firm knows about its business that the regulator does not. But uncertainty does create shared ignorance of common problems, and the significance of the residual asymmetry dwindles as the importance of the threats to both the regulator and regulated entity increases. Regulation by rule making becomes unwieldy as the conditions under which broad measures are to be applied differentiate rapidly and often unpredictably.

Uncertainty that menaces both the regulator and regulated entity arises from three broad sources. First, the rapid codevelopment of innovative products and services by independent firms linked in complex supply chains can introduce latent hazards. Outbreaks of food-borne illness frequently originate in these conditions, for example; so too do potentially lethal defects in automobile airbags. Second, latent hazards can result from the long use of a product or facility under harsh and varying conditions that could not be fully anticipated when it was admitted to the market, as in catastrophes involving nuclear power generation, pharmaceuticals, deep-sea oil platforms, commercial aircraft, or the vast damage done by the long-term application of pesticides, herbicides, and fertilizers.[17] Third, uncertainty can be deliberately created by a regulatory agency, as when it proposes to set a technology-forcing standard, and in doing so, obligates itself and the entities it regulates to extend the frontier of knowledge limits if compliance is to be eventually possible.

In short, uncertainty requires collaboration—typically in the form of a regime that closely monitors developments and sounds the alarm at the first sign that trouble is brewing. In incident or event reporting systems, for instance, failures and anomalies in products or production processes trigger an investigation of whether the events could be the precursors to worse outcomes. These systems also suggest immediate corrective actions and open inquiry into possible design improvements.[18] In a different configuration, such regimes can be used for the close monitoring of ecosystems for the protection of particular species or the environment in general.[19] In the case of technology-forcing regulation, the joint evaluation of possible standards in light of a shared understanding of developments at the technical frontier both encourages a race to innovate (as capable firms compete to shape emerging requirements) and makes broad acceptance of the eventual measures more likely. We encountered such a regime in the Montreal Protocol's TOCs in chapter 2. We will encounter others in chapter 4 and many other places throughout this book, such as setting vehicular emissions standards in California in ways that advance innovation and encouraging important emission-reducing innovations in the DOE's ARPA-E program.

Despite these collaborative elements, regulation is still a contentious business. That is why experimentalist regimes count on a particular suite of incentives—penalty defaults and conditional offers of support—to be described in detail later in this chapter. Penalty defaults loosen the grip of the status quo on firms, and make innovation and improvement more attractive as strategies. But to succeed, they also need the support and implicit incentives provided by collaborative regimes to reduce the risk and guide innovation. Both penalties and collaborative setting are required.

LEGALIZING PROVISIONALITY

Beyond production and regulation, decision-making premised on provisionality requires a new and distinct kind of law—precise and reliable enough to allow coordination in pursuit of joint goals while protecting the parties against the vulnerabilities that cooperation creates, yet not requiring in advance the kinds of commitments to particular actions that uncertainty forecloses. Since the 1990s, contract and administrative law have evolved in this direction. But like the changes in the organization of production, the legal changes have been patchy, and, unsurprisingly, most pronounced in areas like information technology and biotech where the new methods of production have advanced the farthest.

Take contract law first. In a standard contract, the parties specify exactly what is to be exchanged for what, and when. The failure to perform as obligated triggers traditional penalties, and in contracts between sophisticated parties, possible breaches and sanctions can be elaborately detailed as well. Under disruptive uncertainty, by contrast, parties are by definition unable to specify their respective obligations, as they would in a standard contract. Breaches and penalties can't be specified in advance either when it is initially unclear what can be done. Instead they agree, as in experimentalism generally, on broad goals and a regime for exploring the most promising approaches by regularly evaluating the unfolding prospects of success. The regime provides for periodic, joint reviews of progress toward interim targets or milestones as well as procedures for deciding whether to proceed or not, and with what exact aim, along with mechanisms for resolving disagreements. By exchanging this information, the parties clarify the shared goal, and improve their assessments of one another's capacities and reliability.

This kind of information exchange regime is at the heart of a new type of contract, dubbed "contracting for innovation."[20] Such contracts are routine, for example, in collaborations between a small biotech firm that specializes in some aspect of therapeutics or vaccines and a big pharma company that specializes in organizing clinical trials, regulatory approval, and the manufacture and distribution of eventual products. The information-exchange regime is designed to encourage deliberation through the full disclosure of all available information and full ventilation of all sides of a disagreement. Milestone determinations—resulting in decisions about whether to proceed or not with a project—are made by a committee composed of equal numbers of representatives of both firms with full knowledge of the day-to-day developments. Decisions are by unanimous consensus. A doubter can thus demand additional information and explanation simply by withholding assent. At the same time, obstinacy or manipulation will quickly be exposed. In the case of deadlock, the decision is escalated to high managers, who have no direct knowledge of the project and will decide on the basis of the evidence in the record. In this context, managers are likely to look with extreme displeasure on subordinates who hold up decisions and waste time with unfounded objections, and anticipation of this reaction disciplines the use of the—time-limited—veto power. The consensus requirement here is intended to be information or deliberation forcing, not, as in the UN General Assembly, to protect every party against unwanted change.

In time, the operation of the information-exchange regime reshapes the nature of the collaborative relation. As collaboration progresses, the parties' mutual reliance increases. A partner that has not participated in the efforts will scarcely be able to cooperate as fully and effectively as one who has. This increase in mutual reliance increases the costs to each party of switching to an outsider, making the collaboration more and more self-reinforcing, and endowing it with a stability that, under less uncertain conditions, would have been provided by an initial, long-term commitment. In time, mutual reliance generates or activates norms of reciprocity. This trust is as much the result of collaboration as its precondition—just as the precise aims of cooperation are the outcome, not the starting point, of joint efforts.

Looking beyond contract law, administrative law has also seen a progression away from decision by formal rule making toward regulation by guidance. In this alternative framework, an agency tentatively advises private parties and public officials about how it intends to exercise its discretion or interpret its legal authorities, all while anticipating that conditions and obligations will change.[21] Though notice-and-comment or "legislative" rule making had been anticipated in the Administrative Procedure Act (APA) that codified the operation of the New Deal administrative state, the widespread use of the form is relatively new, dating from the late 1960s and early 1970s.[22] As the demand exploded for extensive regulation in areas affecting the whole economy, such as the environment and occupational health and safety, new agencies were empowered to make general, science-based rules premised on extensive public consultation, rather than following the traditional path of challenging particular practices one by one in administrative proceedings. It was through this adjustment to the traditional path that guidance found an opening.

Notice-and-comment rule making is highly formalized, requiring an agency to explain its purposes in great detail, expose its evidence gathering and deliberation to public scrutiny, and explain its reactions to criticism. The judicial elaboration of the original, formal requirements made a demanding process unwieldy and often unworkable. To take just one example, in proposing a new rule, an agency has to elaborate all the arguments in favor of its proposal in advance, knowing that whatever might be learned from the public exposure of its thinking, a reviewing court will consider only those arguments, in their original form, in judging the legitimacy of agency action. By the 1980s, for agencies such as the EPA, the strategy was to consult key stakeholders before announcing a proposed rule, and treating the formal proceeding as a kind of Kabuki stylization of the actual process.[23]

With increasing uncertainty, however, agencies were often unable to marshal in advance the kind of conclusive arguments that were formally required. But reasoned decision-making was not completely stalled; administrators were not left to choose among alternatives by "throwing darts," as some commentators suggested.[24] Rather, agencies pressed ahead by measured action, building regimes and making decisions that were provisional and calculated to lead to better-informed next steps. To make coordination possible without foreclosing the possibility of early and repeated revisions, agencies turned from rules to guidance—another category of decision-making slumbering in the Administrative Procedure Act.

Guidance, in sharp contrast to notice-and-comment rules, can be issued or amended quickly, with little, if any, formal process. Because of this informality, guidance does not have "the force of law" and "power to bind" private parties formally accorded to regulatory rules. Instead, as one scholar puts it, guidance is "only a suggestion—a mere tentative announcement of the agency's current thinking about what to do in individual adjudicatory or enforcement proceedings, not something the agency will follow in an automatic, ironclad manner as it would a legislative rule." Guidance thus not only permits but also demands flexibility: "If a particular individual or firm wants to do something (or wants the agency to do something) that is different than what the guidance suggests, the agency is supposed to give fair consideration to that alternative approach."[25] Similarly, while an agency may choose to depart from its guidance without formal process, it should in principle give a reasoned explanation for such departures.[26] Guidance is therefore a tool for measured action.

The issuance of guidance is now a dominant form of administrative decision-making. In a recent study, leading public and private administrative law practitioners concurred that guidance is "essential to their missions."[27] A former senior Food and Drug Administration official could not "imagine a world without guidance," for example, and according to a current official of the EPA, guidance is "the bread and butter of agency practice."[28] Guidance in its countless forms—advisories, memos, interpretative letters, enforcement manuals, FAQs, or even highlights—is nowhere systematically collected, but a rough estimate is that the number of pages of guidance that agencies produce overshadows "that of actual regulations by a factor of twenty, forty, or even two hundred."[29] In chapter 4, we will see this kind of guidance at work as the EPA grappled with the problem of how to control sulfur emissions from power plants. Its statutory authority was based on the old idea of administrative oversight, but when the EPA was most effective, it operated in guidance mode.

Though guidance has become a dominant form of administration, we should note that it remains controversial. Guidance can certainly be abused. To advance their institutional interests or those of favored constituents, administrators may use guidance too flexibly, escaping the procedural requirements by which policy is normally changed in disregard of the consistency required by the rule of law. A related fear is that because the provisionality of guidance documents makes them difficult to challenge in court, agencies can use guidance to evade not only the preissuance notice-and-comment process but also postissuance judicial review.[30] Over the years, well-established groups of practitioners have proposed working solutions to these problems, but debates continue over the proper limits to the use of guidance and how best to ensure that those limits are respected. In the United States, these debates are entangled with a larger constitutional division between the Left and Right over the legitimacy of the administrative state, and thus they are not likely to end soon.[31]

The practical conclusion we draw is that administrative law's use of avowedly provisional decision-making will continue to expand as it has over the last few decades, even while important questions remain open.

Experimentalism as a Form of Deliberation

As we have seen, experimentalism has become a new kind of organizational principle. But because experimentalist organizations embrace uncertainty, they inevitably invite doubt—doubt about whether the rules they are currently following serve their goals, and whether the goals themselves should be reconsidered. In this section we consider experimentalism as a kind of deliberation, the distinctive form of discussion by which these doubts are clarified.

A working definition of uncertainty is the condition in which experts disagree, and that condition is the starting point of deliberation in experimentalist settings. Between doubts about the facts and doubts about the theory, no one knows for sure. Deliberation takes such disagreement for granted, but it does not devalue expertise. It treats experts not as repositories of fixed wisdom but rather as guides to investigation. Lay experience of problems and possible solutions can be as valuable in such inquiry as the contending views of the experts themselves.

In experimentalism, deliberation is typically organized as peer review, in which actors of equal standing—all with experience of the problem, though of different kinds, and all with a stake in the outcome—evaluate an identical situation, consulting experts as needed. Deliberative peer review does not produce certainty. But it does dispel enough doubt to enable action that both addresses the problem and yields information about how to better the response: measured action.[32] Such deliberation is a deeply collaborative venture in which participants expect their minds to be changed and learn from their differences. In this kind of deliberation, Dewey's shoemakers and shoe wearers get down to cases. Expertise is neither venerated nor vilified but instead seen as the indispensable resource it is.

Though deliberation has a venerable history in both political theory and democratic practice, it has recently come in for heavy weather. Amid ever-spiraling political polarization and a crisis of disinformation, many have raised doubts about the possibility of deliberation, not only among the public at large, but within hallowed institutions like the US Senate specifically designed for that purpose. This skepticism finds implicit endorsement in a large body of recent academic work that focuses on the irrationality of individual decision-making—ignoring institutionalized reasoning as a corrective or substitute for individual choices.[33] Indeed, where group decision-making is mentioned in this literature, it is mostly linked to the dangers of groupthink, in which the pressure to conform compromises deliberation and may even open the way to extremism.[34]

Against this backdrop, our argument that deliberation is key to decision-making under uncertainty may seem like magical thinking. But while there are grounds for this skepticism, there are also good reasons to resist it. In fact, doubts about the possibility of deliberation in public life have prompted new research, coupled with demonstration projects, suggesting that deliberation on the lines of peer review is far easier to organize than concerns with polarization may lead us to expect. In studies of "deliberative polls," for example, a group of some 125 citizens, randomly selected to mirror the electorate as a whole, is provided with curated, balanced materials on a particular, salient controversy. Participants spend several days considering the topic in randomly assigned small groups, interspersed with plenary sessions in which participants engage experts on questions that arise in group discussion.[35] The back and forth in deliberative polls between information gathering and reason giving about the implications of what is found closely approximates the process of peer review, with the qualification that there is more deference to experts as guarantors of sound, uncontroversial thinking in the polls than in peer reviews. Deliberation is convened precisely because expertise alone has hit limits.

The results of these studies are encouraging. Surveys of participants' views before and after the process regularly show improved understanding of the issue along with considered shifts in views that belie the effect of groupthink.[36] However much one doubts that these deliberative processes can be massively scaled up to transform electoral democracy, as their most determined supporters suggest, the success of deliberative polling in many countries shows that under broadly favorable conditions of the kind readily available in experimentalist settings, deliberation comes naturally.[37]

To illustrate how deliberative peer review can take root in practice, we turn now to a case study by social scientist and law professor Daniel Ho that explicitly tests the performance of experimentalist peer review in the reform of a food safety program facing significant challenges in Seattle and King County, Washington.[38]

A CASE STUDY IN PEER REVIEW

The problem of frontline discretion is inherent in bureaucracy and high on the list of attributes said to be its ruin. Rules are made at the top of hierarchies, but interpreted and applied at the bottom. It is the immigration judge deciding on an application for asylum or administrative law judge reviewing a case for Social Security disability benefits who determines what the rule maker intended. If anything, there is even more room for discretion when frontline decisions are taken informally, on the spot. The teacher behind the closed classroom door or police officer on the beat—sometimes called street-level bureaucrats—make nearly unobserved decisions deeply affecting the life chances of those before them. The obvious remedies are more detailed rules or fewer but unconditional ones. Both fail. Multiplying rules only leads to conflict among them and invites further discretion, while mandatory and uniform requirements, applied without regard to context, can have disastrous consequences, as in the case of mandatory sentencing requirements in some US criminal proceedings. When people give up on bureaucracy and large organizations generally as a way of solving any but the most routine problems, the apparently intractable governance problem of frontline discretion—high-level principals can't effectively direct the action of low-level agents—is often the reason why.

In 2014, just before the Seattle experiment, the city's food inspection department, operated jointly with King County, was this kind of street-level bureaucracy. Inspection styles varied from pedagogical to punitive. One of the fifty-five inspectors assigned an average of 1.8 "red points" per inspection—signaling a critical violation, such as failure to wash one's hands, thereby increasing the risk of food-borne illness. At the other extreme, ten inspectors assigned on average more than 10 red points.[39] The rotation of inspectors to new areas had little effect on the outcomes; the person, not the place, was determinative. Some colleagues filed grievances about the practices of others or went on TV to accuse the department of willfully ignoring violations in ethnic restaurants.

Food inspection in Seattle was inaccurately lenient as well as inconsistent. In neighboring Pierce County, county inspectors (more focused on quality assurance) and Food and Drug Administration inspectors found more violations in chain restaurants than were found in restaurants of those same chains in Seattle. Since franchisors (whose reputation depends on the consistency of the customer's experience) go to great lengths to ensure that franchisees maintain standards, faults were most likely as frequent in Seattle as in Pierce, but Seattle inspectors found them less often. A broad review of the food inspection program in 2014 brought the underlying weaknesses of the system to light and prompted Seattle, aware of Ho's earlier work on food inspection, to ask for a menu of options, from which the city then chose the introduction of experimentalist peer review with a randomized control to allow for the careful evaluation of results.

The design was to have two inspectors—the peers—spend one day a week together independently evaluating restaurants or other establishments at the highest level of risk for food-borne illnesses, and adjusting their understanding of what they had seen through a discussion of disagreements after each inspection and over lunch—that is, via deliberation. Pairing inspectors eliminated the possibility that divergent judgments reflected different facts. Supervisors as well as frontline workers were included in the experiment to emphasize, as experimentalism envisages, that reform was to encourage learning throughout the organization and allay any suspicion that the true purpose of the initiative was to increase management's control of the frontline staff.[40]

The original reform design was modified a few weeks after the start of the experiment on the discovery that disagreements were even more frequent than anticipated—peer reviewers were disagreeing on the code implementation 60 percent of the time—and that the disagreements often involved thorny, red point questions that pairs could not resolve in discussions. Weekly "huddles," originally intended as brief meetings to resolve logistic issues, were expanded into problem-solving sessions to consider hard cases.[41] The huddles included all the frontline inspectors and various levels of managers participating in the peer review experiment.[42] They collaboratively tackled questions of code interpretation, occasionally seeking additional advice about food science and law from Ho's team at Stanford or about specialized topics like the organization of outbreak investigation from outside experts.

While a few questions could be resolved solely by a close analysis of the Food Code, ambiguities in words were often linked to ambiguities of substance. Was vacuum-packed meat, sealed in a plastic bag, sufficiently separated from food stored immediately below? What about shelled eggs in the same position? In the huddles, the peers worked toward resolutions of these ambiguities by evaluating their own experiences with the support of the teams' investigations. Pictures of actual cases were frequently used to keep the particulars of problems clearly in view.

The huddles proved to be the engine of deliberation in the experimentalist reform, providing the kind of forum for discussion of concrete differences among informed and engaged participants, in consultation with experts, in which minds are apt to change. The consistency of inspection judgments improved as a direct result of huddle deliberation. The rate at which peers disagreed on red meat violations, for instance, fell almost to zero during the weeks in which the huddles focused intensively on food contamination. The accuracy of inspections improved as well. King County scores converged with those of Pierce County, which had already undergone more extensive quality assurance efforts. Tellingly, the convergence was driven by the greater willingness to award red points on the part of inspectors habitually hesitant to do so—members of the "lenient" group in King County.[43]

The huddles changed the nature of the food inspectorate as an organization. The organizational change involved a shift in the understanding of inspection from a "checklist" approach in which the frontline worker decides they are not seeing a code violation, to risk assessment, in which the staff, together with outside experts, establish the conditions to consider in judging whether a particular situation violates safety requirements. An encounter with taro root, a staple in the African, South Asian, and Oceanic diets only recently introduced though ethnic restaurants to Seattle, encapsulated the motive for the change. The checklist question was whether to class taro as a "potentially hazardous food" whose temperature has to be controlled to avoid the growth of microorganisms. But investigation showed that the risk of taro becoming contaminated also depends on moisture and the pH. A temperature requirement alone would be insufficient; in isolation, it could be dangerously misleading. The risk assessment answer to the binary choice posed by the code was instead to identify the risk factors and check that the establishment gives them due consideration.[44]

The new understanding of inspection as risk assessment also led naturally to an increasing reliance on guidance as the form of regulation: peer reviews call attention to consequential conflicts in code rules; the huddles, in collaboration with the specialist teams, elaborate risk-based clarifications and translate these, as part and parcel of developing training modules for the issue, into guidance documents; and inspectors embrace the guidance because it incorporates their own experience to surmount the difficulties they have faced.

PEER REVIEW, DELIBERATION, AND ORGANIZATION

This case study in Seattle illustrates four key aspects of the mutually supportive relation between experimentalist organization and experimentalist deliberation.

First, the introduction of peer review turns hierarchies into experimentalist organizations that are neither top-down nor bottom-up. With peer review, proposals to change the rules or their interpretation can come from either the frontline inspectors or supervisors. There is no clear superior. As bureaucracy gives way to continuing exchange across levels of the organization, the governance problem of street-level bureaucracy becomes tractable. When frontline workers collaborate in revising the rules, they exercise discretion in the open, justifying their choices with good arguments.

Second, deliberative peer review depends on the continuous interplay of disciplined investigation and reasoned interpretation of what it brings to light. Until huddles allowed consultation with experts, disagreements among inspectors, many with years of experience, went nowhere. Consultation resolved disagreement, though often by recognizing the need for risk assessment, which entails continuing inquiry into possibly dangerous circumstances and what to do about them. In this setting, experts do not crush deliberation with pretensions to technocratic authority. Rather, they enable it.[45]

Third, there is a feedback loop: experimentalist organization encourages deliberation, which in turn encourages further experimentalist reform. The upshot is an institutionalization of provisional decision-making. Introducing the peer review of inspections exposed disagreements whose persistence, having become routine, had hidden defects in decision-making from view. Coming to grips with those disagreements drew attention to the defects; grappling with defects led to deliberation in the huddles; that deliberation gave rise to new routines in inspection and training, and a corresponding shift to administration by guidance in recognition of the need to regularly revisit practices.

Finally, peer review revealed a capacity for self-determination on the part of frontline workers that belies widespread skepticism about deliberation. Long-serving workers, unionized in the public sector, with job security and a history of internal strife, might not be expected to change their ways. But they did. Where the frontline workers saw traditional quality control as "top down," they saw peer review as "bottom up; we get our hands greasy." It helped them reconnect to their profession and peers. Huddles allowed open discussion of complex and controversial distributional issues, such as whether particular regulations unduly burdened ethnic restaurants, which had been talked about furtively or not at all. Ho is emphatic that "the group dynamic of peer review may be a particularly effective way to unsettle and disrupt longstanding habits. Individuals can be more open to change than one might think."[46]

Much of this book will be concerned with frontline workers—those who put policies into practice in making goals concrete and operational. But we will use the term in an expanded sense. As uncertainty increases, goods and services are produced more and more collaboratively, often with the participation of those who use them, and the scope of frontline work increases. In a medical setting, for example, the frontline workers or street-level bureaucrats are traditionally doctors and nurses. Today the group would include patients and patient support groups as well, as they will frequently participate in elaborating new therapies, or testing the safety and effectiveness of new drugs. This difference in the scope of the term aside, frontline workers, however subordinate they may be in society or their organization, are expert in some aspect of the problems that concern them. That expertise may be distinct from, and in some ways more limited than, the expertise of the superiors authorized to make rules. Yet it is often key to achieving the desired outcome. If the account here holds true, as uncertainty and the impossibility of specifying outcomes in advance generate pressure for collaboration, bureaucracy will dwindle, and with it the importance of street-level bureaucrats. Frontline discretion exercised by peer review, though, will become all the more important.

Experimentalism as a Set of Incentives

We have now explored experimentalism as both an organizational structure and form of deliberation. In this final section, we consider experimentalism as a distinctive set of incentives for experimentalist action.

To see the necessity for incentives, consider the following thought experiment. Suppose that provisional decision-making can be institutionalized and given legal form, and that within these structures, deliberation makes disagreement under uncertainty a wellspring of progress. These structures and routines force consideration of how things could be different—or more bluntly, what is wrong with the way things are now. This process is intrusive and disruptive by its nature, however much it aims at improvements that will ultimately lighten the load. Even if lead actors are already engaged in related activities in some domains, they may hesitate to extend the scope of that engagement to climate change or other areas of public concern. Laggards, already struggling to acquire needed capacities or resigned to doing without them, may reject new responsibilities out of hand. In short, even in a world where the background capabilities for experimentalist governance are increasingly recognized as essential, many actors—perhaps most—will prefer the status quo to new and demanding regulatory responsibilities, and the wavering or outright opposition of some will discourage participation by others.

A penalty default is a sanction designed to break the grip of the status quo and encourage participation in experimentalist governance problem-solving when the public interest requires it, but immediate self-interest does not. The most common penalty default is exclusion from a market by denial of a license or certificate of conformity with standards, or by regulations. But the sanction could include loss of decision-making control to an outside authority or other draconian punishment—*in terrorem*, as lawyers say. Such penalties so limit freedom of action that actors will almost always prefer to work toward a feasible alternative, however uncertain initially, that reflects their preferences than to suffer the sanction. Penalty defaults make it risky for capable actors to fail to make good faith efforts to achieve demanding results and less competent ones to fail to make any progress at all toward improvement goals, once they have been set. At the same time, penalty defaults do not sanction the failure to meet targets whose feasibility was unknowable at the outset, nor the good faith efforts of laggards to improve, even when the goals of improvement are not fully met.

To discourage obstruction while encouraging cooperation and improvement, penalty defaults combine an unconventional kind of penalty applied as an unconventional kind of default rule. Conventional penalties are designed for a world of risk rather than one of uncertainty. In this world, actors know the odds and are fully capable of carrying out whatever decision they make; the actors are, at least for the range of decisions currently before them, omniscient and omnipotent. Under these conditions, a penalty is set high enough so that the costs of violation (discounted by the probability of detection) just outweigh the gains from breaking the rule.

But this model is inappropriate in a world of high uncertainty and complexity. When the goal is to do the previously impossible, a comparison of the costs of compliance versus violation is meaningless as neither can be calculated. Even when the goals do become more settled, though, the failures of individual actors often result from incapacity, not from a calculation of costs and benefits. A firm or country making its best efforts to comply with a standard or regulatory requirement won't have a better idea of what do because it now faces the prospect of a fine, and in many cases can't simply hire an expert to tell it what to do either. In short, the usual incentives for fully informed and capable actors don't work when actors face uncertainty and the limits of their own competence.

In place of a menu of differentiated penalties that make departures from rules cost more than they are worth, a penalty default substitutes a binary choice acknowledging the uncertainty of the situation: pursue the best, feasible alternative to the status quo given your situation, with support if necessary, or invite an outcome that will be prohibitively costly and burdensome. For capable actors, who stand to benefit as front-runners from advances that are incorporated into standards or regulations, the best alternative to the status quo is likely to be the pursuit of improvement in the desired direction, frequently in collaboration with others similarly situated. For laggards, the best, feasible alternative to the status quo is likely to be a renewed effort to catch up, but with the help of various technical and financial support programs.

Applying such sanctions as a default is also unconventional. Conventionally, a default is the rule that a judge or other actor imposes when the parties to an agreement have made no provision of their own for certain circumstances. In that case, the judge applies or devises the rule that maximizes the parties' joint welfare on the assumption that this is what they would have done had they attended to the matter with the information as well as capabilities they are reasonably presumed to command.

A penalty default refers to the special case where the judge—say, in divorce proceedings—doesn't have the information to devise the optimum rule, but the parties most likely do, even if they are reluctant to provide it.[47] To induce the parties to cooperate, the judge presents a stark choice: agree on a division of assets and income, or the court will impose a settlement that will surely cause great hardship. While deliberation cannot be compelled, the prospect of this penalty makes clear the costs of failing to deliberate and incentivizes the parties to consider using what they know of their situation to devise a solution no outsider could anticipate.

In the same way, experimentalist governance penalty defaults induce reluctant parties to elaborate missing, default terms of cooperation, but with two differences. First, under uncertainty, even the most capable actors, alone or together, are far from omniscient in the sense supposed by theories of decision-making under risk. They will not know enough to adequately identify and gauge their choices, and that done, how to realize the ones they prefer. They will need to supplement what they do know by learning more through investigation, pooling efforts where interests allow; often this investigation will be too costly and risky for private actors to undertake without public support and facilitation. Second, there can be problems of capacity. Many actors are not omnipotent in the sense supposed by traditional decision theory either. Laggard firms, as noted, will need support in acquiring the capacities for self-monitoring that compliance with the new requirements demands.

Experimentalist governance penalty defaults are typically necessary, therefore, but in the absence of support may not be sufficient to loosen the hold of the status quo. Since the entities that provide these elements of a problem-solving regime may not be the same as those that can credibly threaten draconian penalties, the application of penalty defaults in experimentalist governance may involve coordination problems that don't arise when the threat of an unworkable outcome is enough to get the parties to cooperate in the solution of a problem that was already within their reach.

Setting such problems aside for the moment, we can draw an important conclusion. The commitment to a demanding rule combined with the promise of continuous consultation, all under the threat of a penalty default, shifts the preferences of leading and lagging actors.

For leaders, this process creates an incentive to engage in a broad exploration of possibilities. Given the regulator's commitment to act, inertia no longer favors the status quo. Since a rule is coming, the actors with the most confidence in their improvement strategies consult with the regulator in an effort to have their preferences incorporated into the rule, minimizing their own costs of adjustment and raising the costs to competitors with different approaches. Given vertical disintegration, suppliers specializing in the sought-after solutions will be among the most prompt and persistent volunteers. Their business model is indeed demonstrating the feasibility of the kinds of solutions they pioneer. And the prospect that at least one actor will cooperate with the regulator induces others to cooperate as well, both to secure a hearing for their own solutions and learn, through cooperation in the various review groups that the regulator establishes, what competitors are up to.

This broad participation ensures in turn that the regulator's decision is informed by good estimates of short- and medium-term possibilities, and corrected as efforts at implementation warrant. Rules and revisions thus result from joint learning among actors, none of whom could devise a solution alone.

Laggards, for their part, will likely have kept to the sidelines as the standards are set. Once they see that compliance is feasible, many will consider bringing themselves into line, but will have trouble implementing and perhaps even fully understanding the adjustments that are needed. Imposing penalties for such violations, as though they were calculated or in bad faith, can have perverse effects. Actors who fail to meet obligations out of ignorance or incapacity may be driven to conceal shortcomings from authorities, inviting duplicity where there had only been good intentions. Penalty defaults, which eventually insist on compliance with regulatory standards, are typically forgiving in such situations, treating initial violations as presumptive evidence of incapacity, and leaving room for training, extension services, and other forms of support to weaker actors to help them learn to meet the requirements. By the same token, there are no penalties for reporting breaches of rules; indeed, timely reporting usually mitigates any eventual liability. Often the regulatory requirements are explicitly adjusted so that distinct and less well-resourced groups of actors—the most common examples are small firms and farms—can meet the necessary standards by procedures suited to their situation, and frequently in stages, over longer time periods than those set for compliance by more resourceful competitors.

But this forbearance and support has limits. Truly incorrigible actors—those that persistently fail to learn or demonstrate that they have no intention of doing do—are eventually subject to the full penalty default: exclusion from the regulatory regime and the associated market by the denial of a necessary permit, conformance certificate, or quality mark. Rents and other resources erode away from firms that persistently fail to adjust—rendering them impotent and then nonexistent. Penalty defaults are forgiving until they aren't.

How do penalty defaults take shape in practice? They arise from many sources, and can be effective whether embodied in formal legal sanctions or not.

Frequently the source is moral outrage, although often the precise trigger for the outrage is unclear. Experiments show that consumers are little inclined to pay a premium for "ethically" produced goods—for example, "sweat-free" T-shirts made under certifiably high labor standards.[48] But the same kind of consumers will often boycott firms caught flagrantly violating environmental or labor norms. International brands with a reputation for respecting these norms are of course particularly vulnerable to such reaction; knowing this, international NGOs have become extremely adept at calling attention to corporate breaches of widely shared moral convictions.

Mobilizing public opinion in this way, Greenpeace, as we will see in chapter 5, was able to establish a monitoring regime for corporate producers of soy and beef along with their suppliers in the Brazilian Amazon, effectively limiting some aspects of deforestation, long before governments and trade standard-setting organizations could begin to achieve similar results.

Smaller companies in local communities, operating under a "social license" dependent on continuing acceptance of their behavior—pulp mills in isolated forest settings, for instance—are also exposed to moral pressure, all the more easily generated and effectively applied by neighbors and employees intimately familiar with the companies' practices.[49] In any case, the countless, successful campaigns—local, national, and international—by NGOs to hold companies accountable for their environmental actions clearly demonstrate that normative concerns generate penalty defaults across a wide range of settings.

A second source of penalty defaults is law. The Endangered Species Act is one US example: listing a species as endangered can stop development in its range. Others are contained in the Clean Water and Clean Air Acts. Under the Clean Water Act, the EPA can stop development surrounding a body of water if the inflow of pollutants exceeds the total maximum daily load.[50] Under the Clean Air Act, the EPA can block development plans in urban areas that persistently fail to meet standards—a penalty so onerous that it has never fully been applied, yet is credible enough to force even the most reluctant cities to act.[51] Development can only proceed if the affected parties establish a mitigation plan acceptable to the regulator. In both cases, the ground-level actors elaborate the actual solution, but are induced to do so only by the certainty that they will lose their autonomy if they do not.

A third source of penalty defaults are asymmetries of power and economic position. It is to the advantage of powerful (or simply capable) actors to impose standards, rules, or codes of conduct on themselves and weaker ones alike so that customers, citizens, or the world can see what separates them, and reward the well intentioned and high performing. A commercial illustration can be found in private phytosanitary and other quality standards (such as GLOBALG.A.P.) imposed by wholesalers or large retailers on producers of meat, leafy greens, or vegetables connected to global supply chains. The "California effect" to be discussed in connection with CARB in the next chapter and the corresponding "Brussels effect" of the European Union make market access contingent on compliance with "domestic" environmental regulation to set regulatory standards for outsiders, and indeed the world.[52] The United States used its market power to protect dolphins (ensnared as the bycatch of tuna fishing in the eastern tropical Pacific) under the Marine Mammal Protection Act—initially by requiring countries exporting to the United States to adopt the same protective measures used by the US fleet.

In practice, penalty defaults can frequently arise from several of these sources in conjunction. The California and Brussels effects as well as the dolphin protection legislation and Montreal Protocol all wield economic and political power to back legal authority. Often the various sources of legitimacy are invoked in sequence. Moral pressure can lead a large firm to join with other producers and stakeholders in a roundtable to establish a code of conduct, including environmental and labor standards, binding on the whole supply chain. Public authorities can then make compliance with (some of) the provisions of the code a condition of access to the domestic market, thereby obligating foreign producers as well, and changes in the "private" codes are likely to quickly effect public laws—further blurring the distinction between them.

In sum, penalty defaults, though indispensable to experimentalist governance, are only part of the story, and generally the less important part. Unless provisional decision-making by deliberation—measured action—comes to be routine, penalties will be insufficient to induce learning and improvement. In a world where the actors are neither omniscient nor omnipotent, it would be foolhardy to assume otherwise.

Uncertainty as Routine

The combination of these distinctive organizational structures, forms of deliberation, and incentives, all adapted to uncertainty, provides the foundation for a new era of experimentalist governance.

In hindsight, we can see that the price shocks of the 1970s were only the beginning of an epochal shift. Since then, uncertainty and complexity have not only persisted but also increased—especially in the industries that will be central to addressing global climate change. Firms and regulators, we have argued, have come to accept both as a constitutive circumstance of decision-making. Firms in advanced sectors are already practiced in collaborative design with suppliers and forms of production based on short learning cycles and deliberate exposure to vulnerability. Regulators are shifting from the promulgation of fixed rules to the issuance of guidance that invites correction. The idea that initial decisions must be treated as preliminary and subject to revision in light of experience is becoming institutionalized and legalized. Actors revising projects and regulations under uncertainty need to deliberate, and they can learn to do so rapidly, even when they have spent years assessing compliance by checklist in a bureaucracy.

Many organizations, public and private, are adopting these methods under the press of events—because their customers or clients demand it, or because competitors plan to do so. Still, diffusion is patchy, as the merest glance at the gap between advanced and stagnant sectors in the US economy confirms. Where the public interest demands it, incentives can encourage an embrace of the new methods when self-interest doesn't. In short, a world of shocks and surprises obligates us to find a way through uncertainty by learning from the unexpected, and better equips us to respond to climate change in the bargain.

In the next chapter, we show that institutions with these capacities are at the core of some of the most successful responses to major environmental challenges, both in the United States and globally.

4

Innovation at the Technological Frontier

THREE POLICY ICONS AND A COMMON APPROACH TO UNCERTAINTY

We have argued that deep decarbonization depends on innovation at the technological frontier. In this chapter, we look at three iconic cases of successful frontier innovation. We first examine the success of the California Air Resources Board (CARB) in orchestrating deep reductions in transportation emissions by tightening standards for combustion engines—a success that laid a foundation for the state's zero-emissions vehicle mandate of the 1990s. Second, we consider the development of scrubbers, a technology for controlling SO_2, the leading cause of acid rain and other environmental perils, in the context of a pioneering cap-and-trade system of pollution permits. Last, we discuss radical clean energy innovation through the Advanced Research Projects Agency (ARPA-E) of the US Department of Energy (DOE).

These examples cover the gamut of policy strategies that have also been proposed as approaches to the climate crisis. The case of ARPA-E is typically taken to illustrate the view that markets are too preoccupied with short-term calculations to invest in radical innovation; for that the state must step in. The case of scrubber development is often taken to point in the opposite direction: the solution to complex environmental problems is beyond the reach of government bureaucracy; only markets are fit for the task. The success of CARB is taken to confirm a middle ground position: empowered by the ability to threaten the closure of a large (state) market, the descendant of the New Deal administrative state can set rules to force change.

There is something to say for each of these strategies as a way of grappling with market failures. But we contend that when it comes to the mechanics of *any* of these approaches—how to ensure that the state picks truly promising research areas and projects; how to identify feasible goals for technology-forcing programs and correct errors when they occur; and how to create the right background conditions for the market, ranging from trading rules to technological innovations that the market will not produce itself—experimentalist governance is indispensable as a means of addressing the uncertainties that frustrate conventional approaches. To ignore this need for experimentalism is to run the risk of repeating the same errors that turned the promising Montreal model into the failed Montreal mold.

All three of the cases we examine are from the United States. This is not for want of instances from elsewhere; there are plenty in the literature, and chapter 5 discusses some of them. But even those who are most open to programmatic innovation might cast a dubious eye on the relevance of foreign examples. Could it really happen here? What about the US reflex—sometimes justified, and sometimes not—to see collaboration with industry as regulatory capture rather than a path to problem-solving? And what about the baroque and ossified formalism of US administrative law that allows—and perhaps even encourages—reluctant firms to challenge new standards in court with good chances of success? With three powerful homegrown examples, we show that these objections can't be as robust as frequently assumed—just as the examples in chapter 3, where we outlined the theory of experimentalism, demonstrate that this mode of governance is highly effective under the right conditions. The space for innovative collaboration and experimentalist governance is a lot greater than assumed, even in the United States.

What Was Miraculous in the California Miracle?

Our first case begins with smog, an unhealthy urban pollutant primarily the result of combustion emissions from motor vehicles and industrial operations. Since the 1960s, California—whose cities had (and still have) some of the worst air quality in the nation—has pioneered successful regulatory efforts to limit the emissions of smog precursors and other important vehicular pollutants. Because of the state's trailblazer status, the 1970 Clean Air Act allows California, alone among the states, to set its own emissions standards, provided they are at least as protective as the national ones; under the statute, other states may (and in fact many states do) adopt the (unaltered) California standard.

By these mechanisms—demonstrable leadership in its home market and easy followership by others—California has set the pace, especially in recent decades, for the reduction of vehicular emissions in the nation as a whole. Smog levels in California have dropped by 60 percent since the 1960s, even though the number of cars on the road has doubled.[1] Today, thirteen other states plus the District of Columbia have adopted California's standards; about a quarter of all the vehicle miles traveled in the nation occur in states that follow California's rules.[2]

CARB, which has regulatory authority for protecting the quality of the state's air, is governed by a board of twelve members, appointed by the governor, plus two legislators (one each from the state's assembly and senate). By statute, its board must include at least one member with experience in automotive engineering, one with training in chemistry or meteorology, one with expertise in medicine or health, one with experience in air pollution, and two representatives of the general public. Each of the remaining six members represents one of the state's air pollution control districts.[3] By design, CARB thus reflects views from a variety of stakeholders—informed by technical expertise—and creates a forum where lessons learned in one control district can be rapidly transferred to others.

Here we focus on one particularly iconic period in CARB's history: the effort in the 1990s to create a market for zero-emissions vehicles. CARB took that goal to entail the development of electric vehicles, but it soon learned that the technology was not ready at the time. How it responded to that reality reveals why CARB works so well in its core regulatory mission: catalyzing and steering efforts to clean up California's air. We tend to think of electric vehicles as an innovation of today rather than three decades ago—in part because electric vehicle technologies (notably batteries) are presently so much more advanced—yet it was CARB's 1990s' regulatory innovations that redefined ultraclean vehicle requirements and set the stage not just for the current revolution but also for extraordinary improvements in pollution control for internal combustion engines. In effect, CARB confronted exactly the same challenge that the Montreal Protocol parties faced when they sought to cut ODS to zero. They knew the ambitious goals they wanted to achieve, and among at least some of the parties, there were powerful incentives for success. But nobody knew how to get there.

In 1990, partly in response to amendments to the Clean Air Act in that year, CARB introduced the low-emission vehicle (LEV) and zero-emission vehicle (ZEV) programs. Together they established the nation's most stringent standards for the reduction of exhaust emissions from gasoline and diesel vehicles of smog precursors, such as nitrogen oxide (NOx), nonmethane organic gas, and carbon monoxide.[4] The standards in both programs were formally amended a number of times (the LEV standards in 1999 and 2012, for example) and continuously revised between formal amendments. The Clean Air Act set extreme punishments for states that failed to clean their air—at the limit, the federal EPA could impose onerous restrictions on economic activity—but those vague punishments were conditional on effort. States had to show, through State Implementation Plans, that they were making progress; the dirtiest states had to demonstrate the most effort. Moreover, inside California, state political leaders were aligned around cleaning the air and faced electoral punishment if they failed to make efforts. Penalty defaults were numerous.

Both of CARB's vehicle programs induced and codified major changes in technology and emissions. The LEV program's ambitious standards drastically reduced smog emissions by inducing incremental innovations to the gasoline- or diesel-burning internal combustion engine that have had, over time, cumulatively radical effects. A new car sold today in California emits 99 percent fewer pollutants than a new car in the 1960s. The ZEV program—designed to encourage volume production of cars with "no exhaust of evaporative emissions of any regulated pollutant"—helped induce the development of affordable batteries that could meet range and chargeability standards akin to conventional vehicles as well as appeal to consumers—an extraordinarily ambitious goal given the existing technology.[5]

By the standard wisdom, LEVs and ZEVs were classic examples of what CARB does best: identifying a serious problem and moving to fix it through the imposition of strict rules that force changes in industry. It uses science to set goals and then forces industry to align. It demands the seemingly impossible of industry, but because industry needs the California market—as well as other markets that often follow California's lead—it finds a way to make the impossible possible. Having done that, other jurisdictions follow the miracle—in short, the "California effect."[6]

But this characterization is misleadingly incomplete. Access to a crucial market mattered to automakers, to be sure. The threat of exclusion created a powerful penalty default—imposing a price for the failure to cooperate. Yet by themselves, these incentives would no more have guided actors to workable solutions than the incentives created by commitments and sanctions under the Montreal Protocol would have yielded results without the guidance provided by the TOCs. California's success hinged on the ability of regulators to couple regulatory standards to the ongoing evaluation of critical and rapidly changing technologies, such as pollution control equipment and electric vehicles. In other words, penalty defaults and strict standards are necessary but not sufficient conditions; to work, they need to be applied against a background of continuous adjustment and review. In practice, the LEV and ZEV programs succeeded because, as a leading commentator put it, the regulator was "committed to a continuous process of implementation oversight and regulatory review," and was correspondingly willing "to adjust and change the basic program."[7] When one looks closely at the history, the outcome of this collaboration between regulators and industry was extremely different from what CARB imagined would be the case in 1990. As with essentially all complex regulatory programs, the more ambitious the effort, the harder it is to know ex ante what is achievable.

In this crucial function—linking standard setting to the continuing investigation of technological development—CARB obtained its broad authority by the feasibility requirement in the Clean Air Act's Section 202(a), which specifies that the EPA and other regulators must demonstrate that available technologies provide cost-effective ways of meeting any proposed standard. A line of court decisions, including *International Harvester v. Ruckelshaus* and especially *NRDC v. EPA*, suggested that it was the long lead-in times for new requirements that assured feasibility because they made it possible to correct unworkable rules.[8] "The time element in the EPA's prediction," the court in *NRDC* acknowledged, "introduces uncertainties in the agency's judgment that render the judgment subject to attack. [But] the presence of substantial lead time for development before manufacturers will have to commit themselves to mass production of a chosen prototype gives the agency greater leeway to modify its standards if the actual future course of technology diverges from expectation."[9]

The need to link standard setting to active engagement with technological development was also reinforced by the widespread recognition within CARB, from the outset, of the uncertainty of the overall situation and hence the need to learn rapidly to stay abreast of possibilities. A December 1995 briefing to CARB, *Making ZEV Policy Despite Uncertainty* by the RAND Institute for Civil Justice, put this directly: "Enormous uncertainty and risk suggest a search for near-term policies that, a) enable learning, b) are not susceptible to disaster, and c) can be tailored as new information is obtained."[10]

Learning and the corollary policy updating were central to those within CARB from the early days of the LEV and ZEV mandates.

Another important element of effective collaboration between the regulator and regulated entity was greatly facilitated by deep changes in the organization of the automobile industry. Until the 1980s, the major automobile companies were vertically integrated. They developed and produced key components in-house in order to avoid the risk that a powerful supplier could withhold delivery of a crucial component unless the supply contracted was renegotiated in its favor. But as markets became more volatile and the direction of technological development more uncertain, the risks of owning captive suppliers increased dramatically; a shift in the trajectory of development could make in-house capacity irrelevant, while a shift in the level or composition of demand could make it superfluous. Beginning in the 1990s, automakers divested internal suppliers, and purchased more and more components and subsystems from independent manufacturers. Technological innovation also shifted—away from the automakers themselves, and toward coproduction between the supplier and buyer.

An unintended side effect of this restructuring was to shift the balance of power in relations between the regulator and firms in the industry. Vertical disintegration created important new actors, primarily equipment suppliers, with interests in reducing the levels of polluting emissions, even if this improvement was in tension with the final assemblers' goal of minimizing the overall costs. These interests arose because emissions requirements meant greater demand for vehicle components and thus more opportunity for suppliers—especially for suppliers with the greatest capacity to connect innovation to products. Thus component manufacturers and suppliers frequently approached CARB to pitch new emissions control technologies. By demonstrating the superior performance of their products, they influenced emergent standards and created markets for their innovations.[11]

This reconstitution of interests reduced the information asymmetry that traditionally advantages producers in negotiations with relatively ignorant regulators. Indeed, suppliers and component manufacturers developed many of the newer emissions control technologies, such as turbochargers and electric superchargers. The first LEV standards, in the late 1990s, were shaped by the demonstration of an electrically heated catalyst that drastically reduced emissions.[12] Over this period, CARB did not conduct its feasibility assessments in a vacuum; rather, the assessments were more and more the result of near-constant conversations with both component manufacturers and automakers. Since virtually all the automakers belonged to trade associations that lobby regulators and raise the specter of oligopolistic coordination across the industry, CARB emphasized communications with the automakers themselves, and it regularly recruited experienced engineers from the auto industry to build its capacity to engage in detailed technical dialogue with component and carmakers.[13]

This dynamic atmosphere of innovation and oversight produced standards that were technology forcing without being technology determining. Continuing exchanges between firms and regulators set performance standards with an eye to what was feasible, but those standards did not mandate the use of particular technologies to achieve them. Once the feasibility of a certain reduction in emissions had been demonstrated, each firm was free to find its lowest-cost path to the required result that met a particular standard—for example, the quantity of volatile organic compounds emitted per volume of fuel consumed or mile driven. Indeed, automakers usually devised their own means for meeting new standards. For instance, although CARB staff projected "that gasoline vehicles meeting the more stringent [LEV] standards would require the use of emerging new technologies such as electrically-heated catalysts (EHCs) and heated fuel preparation systems," only one manufacturer ever used an electrically heated catalyst; others used (and further developed) their expertise in systems integration to realize "improvements in combustion control in the engine itself to reduce engine-out emissions."[14] The real contribution of the joint exploration of new technological possibilities, in other words, was to demonstrate that daunting, technology-forcing standards could be met—not to determine, and still less to impose, required solutions.[15]

The frequency and scope of these technological inquiries varies according to circumstance. The general rule is that the more uncertain the technological foundation of a proposed standard, the more extensive and frequent the review—and the more interdependent the technological changes, the more frequent the review of the whole system's potential performance. CARB subscribed to this philosophy, and other institutions that help orchestrate innovation at the frontier do the same—from the technical assessment panels of Montreal to ARPA-E. When CARB initially proposed amendments to the first LEV standards in 1996, the agency produced a staff report that identified "four basic aspects of current emission control systems that vehicle manufacturers have been improving to achieve low-emissions levels, [namely] more precise fuel control, better fuel automation and delivery, improved catalytic converter performance and reduced base engine-out emissions levels." The report listed nineteen potential low-emission technologies that would become available when the new standards took effect, and predicted which technologies manufacturers would use to meet the new requirements.[16]

As an illustration of this continuous, joint assessment, consider CARB's biennial review of the ongoing technological feasibility of the ZEV program targets.[17] These reviews, and the related assessments to which they gave rise, proved essential to the effective pursuit of the ZEV mandate despite the uncertain technological developments on which it depended.

The first review, conducted in 1992 under the pressure of various procedural deadlines, was limited to perfunctory approval of the initial ZEV requirements.[18] CARB staff members continued consultations with emission control suppliers and the volume vehicle producers, conducted their own tests of components and prototype vehicles, and reviewed the findings of the United States Advanced Battery Consortium, a collaborative research project supported by the large auto manufacturers.

At the second biennial review, CARB again found that the LEV and ZEV requirements were feasible and cost-effective but extended the review process in two ways.[19] First, given the centrality of battery development to the ZEV program, CARB established an independent Battery Technology Advisory Panel to assess candidate technologies through visits and follow-on discussions with leading developers of advanced batteries and their customers.[20] Second, in response to questions raised during the review, CARB staff members organized a series of public workshops and other forums, from May to November 1995, in which interested parties ranging from electric utilities to environmental groups and auto manufacturers discussed key ZEV issues, such as electric vehicle infrastructure and the marketability of ZEVs. In the course of these discussions, proposals to modify the ZEV mandate were introduced, and CARB convened a public meeting to consider these and other possible modifications in anticipation of the third biennial review the following year.[21]

On the basis of these consultations and a review by the Battery Technology Advisory Panel, CARB concluded that ZEV technology would not be available in time to meet the 1998 requirements.[22] Battery prices had not fallen as quickly as expected; the industry feared that the premature introduction of ZEV models at high prices and their potentially unreliable long-term performance would produce consumer resistance, thereby complicating the future sales of ZEVs.[23] Acknowledging these concerns, CARB removed the ZEV requirements for the 1998 through 2002 model years, but left the mandate of a 10 percent share of ZEVs in the 2003 fleet in place.[24] It did not give up the goal completely, on the theory that the underlying technological conditions that made prompt change impractical might shift in the future.

(For reference, only in 2019 did electric vehicles of all types account for nearly one-tenth of new vehicles in California.)[25]

In exchange for this relaxation of the rules, CARB entered into a memorandum of agreement with each of the large-volume manufacturers. These memorandums bound the manufacturers to continue developing low-emission cars; produce cleaner cars nationwide and participate in advanced technology development and demonstration projects; share propriety development information with CARB; and commit substantial funds to support the United States Advanced Battery Consortium's research. As under the original LEV program, noncompliance was sanctioned by substantial damages.[26] This approach to relaxing one set of rules while keeping others in place helps illustrate how penalty defaults operate best—within an environment of constant adjustment and review. The automakers were failing to meet ZEV rules through no intended fault of their own, but they were doing better at meeting LEV rules than expected. The regulator, suitably informed through peer review, fine-tuned the location of the goalposts and penalty zones.

The 1998 and 2000 biennial reviews recognized that battery development for ZEVs continued to disappoint. Yet LEVs were developing faster than expected. As a result, CARB created new vehicle categories—the partial-zero-emission vehicle (PZEV) in 1998 and advanced technology PZEV. Cars that met the criteria for these new categories—mainly hybrids and plug-in hybrids—were less polluting than any full-scale internal combustion engine vehicles that previously had been qualified to use the state's roadways. CARB therefore allowed manufacturers to meet their ZEV requirements by, for example, counting five PZEVs as equivalent to one ZEV. It formalized such substitutions as the Alternative Compliance Path in the 2003 amendments to the ZEV mandate.[27]

As experience unfolded, CARB kept adjusting the targets so that the regulatory requirements and understanding of technical feasibility changed in tandem. In the 2007 biennial review, CARB delayed increases in the fuel cell requirement, judging the technology insufficiently mature for application. In time, battery technology and control systems caught up with CARB's ambitions—partly because of large and rapidly growing sales of lithium-based batteries for consumer electronics. By 2011, there were seventeen thousand electric car sales, and that number tripled in 2012 to fifty-two thousand. In 2013, there were sixteen ZEV models available from eight auto manufacturers—nine of them purely battery operated. Currently nearly every major automaker has a ZEV for sale in California, and some—for instance, GM—have decided to shift decisively away from internal combustion toward electricity.

As CARB repeatedly shifted rules to reflect its assessment of feasibility, many of the automakers challenged its authority to adopt technology-forcing requirements that they thought infeasible. CARB successfully defended itself by showing that its administrative process engaged the industry and remained aligned with the latest technical information.[28] Far from freezing development, uncertainty about whether the automakers would prevail in their efforts to steer the content of the regulation was a prod to innovation; neither individually nor in a consortium could firms run the risk of halting their own research, while at least some competitors were likely to plunge ahead. (And those that plunged in the wrong direction soon found themselves punished. BMW, for example, pushed hard with fuel cells while CARB interpreted ZEV rules in electric terms. No market for BMW's hydrogen fuel cell cars ever emerged in California.)

This continuing innovation generated new ideas and technologies, which in turn allowed CARB to keep tightening performance standards as well as better defend itself against legal challenges that questioned its authority and technical competence.[29] Similar challenges have been raised about CARB's authority in other domains—such as in the regulation of carbon emissions that affect interstate commerce—and CARB defended itself with a similar playbook that it honed in the 1990s and 2000s: arguments about grounding its work in impartial, technical competence backed by transparent peer review and within the scope of regulatory deference.[30]

As technology has improved over the last decade—this time unexpectedly to favor electric vehicles even as some of the original equipment manufacturers and suppliers expected fuel cells to excel—CARB has changed the rules again. In January 2012, CARB approved a new emissions control program for model year vehicles 2017 to 2025—a new scheme called the Advanced Clean Cars Program that mandates more ZEVs with greater controls of the pollutants that are precursors to smog, including soot, and focuses on cutting emissions of greenhouse gases. The new scheme also includes an overcompliance rule that allows manufacturers who overcomply with some rules to earn credits that can offset a portion of their ZEV requirement from 2018 through 2021. For regulators, this approach not only offers flexibility to manufacturers—a key demand from some in the industry—but also creates incentives for more advanced experimentation that can be used by CARB to reveal places where future rules can be tightened. Supernormal gains from doing better turn the penalty default on its head.

Comparing today with 1990, the ZEV mandate predictably did not operate exactly as originally envisioned. Uncertainty was too profound in 1990 for anyone to know how the system could, would, or should unfold. Precisely because it kept responding to new and often unexpected developments—some even exogenous to the whole auto industry, such as lithium batteries—the program achieved its goals of radical reductions in harmful air pollutants. At first these improvements came mainly through making internal combustion much cleaner; later they relied on hybrids; in time, full electric vehicles have come to play a more central role. Along the way they underwrote the explosive growth of hybrid electric vehicles in the early 2000s.[31] In turn, the development of hybrids has translated back into the substantial development of pure battery electric vehicles in recent years. As of 2010, California had 2.2 million electric vehicles on the road, 80.8 percent of which are PZEVs (mostly hybrids), 17.6 percent are advanced technology PZEVs, and 1.6 percent are ZEVs.[32] And today, those shares are shifting quickly in the direction of full electric models.[33]

From the experimentalist perspective, the ZEV and LEV programs show how rules coevolve between the regulator and supplier. Specific targets for a reduction in ozone precursors and other air pollutants set an initial, high-level goal. CARB served as a central coordinating body for an iterative learning process structured around biennial reviews and many other forums. On the basis of pooled reports discussing extensive field experience implementing the regulatory mandates, CARB periodically—and sometimes substantially—adjusted its requirements. It set rules that forced massive investment in new technology—and competition among suppliers—but did not pick winners. A penalty default—the threat that uncooperative behavior could ultimately result in exclusion from a leading market as well as civil penalties for failure to meet the program's substantive goal—existed at the outset. All of these features, not just any one of them, were essential to make the program a success in pushing forward the technological frontier.

The Misunderstood Miracle of Sulfur Control: Before and after the US Sulfur Emission Trading Program

SO_2, often just called sulfur, is one of the most noxious air pollutants. When the federal government adopted the 1970 Clean Air Act, the first sustained nationwide effort to control air pollution, it was already known that SO_2 was a major source of respiratory disease, an aggravation for asthma, and a cause of numerous other ailments, including in children. SO_2, it would be learned by the 1980s, was also the chief cause of acid rain, which killed forests and other fragile ecosystems. The Clean Air Act created a list of National Ambient Air Quality Standards (NAAQS), which named SO_2 as one of the inaugural six pollutants to be regulated. The EPA, also created in 1970, was charged with keeping the NAAQS up to date and setting standards for "clean air" by looking to the science of public health, without consideration of cost.

At the time, most SO_2 came from coal-fired power plants. The United States has always burned a lot of coal for power, but when the first oil crisis hit, the country increased its dependence on coal, nearly all of which came from US mines; it shut plants fired with costly and insecure oil, and built more coal plants to keep up with the rising demand for electricity. By the 1980s, coal accounted for about 60 percent of all domestically produced electricity. Thus the regulation of SO_2 implicated a well-organized, important, and politically powerful industry: electricity. Controlling sulfur would require accommodating powerful economic interests.

Five decades later, US emissions of SO_2 have fallen tenfold, from 31 million tons annually to about 2.8 million tons today. In the conventional account, this tremendous success was due to the timely recognition that government was not skilled at administrative intervention—or the command-and-control prescription of technology, as it was and still is called. Though it was armed with the NAAQS, a mandate to clear the air, and the authority to impose technology standards under the Clean Air Act, the EPA nonetheless failed to achieve satisfactory results from the 1970s through the 1980s. Those failures had become apparent by the 1990s, when Congress, at the urging of powerful environmental groups and business interests worried about cost, authorized creating a national trading scheme with amendments to the Clean Air Act.

It took five years to get the new market up and running, but once in operation it worked like a charm. Prices for sulfur permits were less than half the level that had been expected. The conventional story attributes this drop to the power of market forces in finding inexpensive solutions that administrative action could not.[34] This success only further entrenched the celebration of markets as heirs to failed bureaucracies such as the Soviet State Planning Committee, known as Gosplan. A retrospective by one of the most active NGOs, the Environmental Defense Fund, is titled "How Economics Solved Acid Rain."[35] Writing in 2002, the *Economist*, looking back at the history, called the sulfur market "the greatest green success story of the past decade."[36] (The 1990s also included the ramp up of the Montreal Protocol, creation of the UNFCCC, signing of the Kyoto Protocol, and Rio Earth Summit—so the pronouncement was a bold one.)

The story we tell here differs dramatically from this standard account. Administrative action was working quite well in forcing needed technological innovation. Without new technology, it would be impossible to make deep cuts in emissions. The breakthrough came with scrubbers—devices that chemically remove sulfur from flue gases before being pulled up a smokestack. The technology was developed and tested from the early 1970s through the 1980s, long before the sulfur market existed. That innovation emerged through collaboration between industry (led by the largest and most motivated firms), the EPA, equipment vendors, and a research institute created by the utility industry known as the Electric Power Research Institute (EPRI). That collaboration mirrored the patterns we observe in CARB. Hardly an effort in heavy-handed command and control, it was a collaborative system that ran multiple experiments in parallel, constantly adjusted means and ends, and chose technologies based on performance.

Ironically, regulation was coming under sustained attack by many US political and business elites just as this experimentalist approach was realizing its most profound benefits of innovation and dynamic efficiency; only by the later 1980s was it clear which scrubber systems would perform best, and whether it would be possible to operate those systems at high reliability and rising efficiency. The sulfur market "worked" in part because a regulatory and innovation system that operated according to experimentalist principles had already supplied the critical innovations that were diffusing into service as more power plants learned which scrubbers to install and how to operate them. When the sulfur market began operations around 1995, it helped allocate resources efficiently in the short term across known technologies—a process often called static efficiency. Even in that respect, however, its contribution was more limited than it appears.

Another major contributor to lower-than-expected sulfur prices was a sharp increase in the availability of less polluting sources of coal—an option that emerged from fortuitous changes in other bodies of regulation not tied to the trading system. Sulfur prices were low, relative to expectations, not because markets were superior to bureaucracy in finding dynamic solutions. Instead, unexpectedly low prices appeared because no analyst at the time anticipated how all of these technological and coal market changes would interact to make pollution control easier than expected.

Although today the sulfur trading system has a hallowed place in the story of the triumph of markets over administration, in fact the period of market success—when prices were lower than expected and stable—lasted only about eight years. So long as there was little need for major technological innovations, and no need to make plant- and location-specific choices, the nationwide market helped achieve modest gains. But this approach failed to achieve the mission of the Clean Air Act: clean air everywhere. A national market allowed pollution hot spots to emerge; places where it was cheap to pollute attracted more polluters.

At the same time, new science identified new pollutants and new local interactions between pollutants, which meant that effective pollution control would require addressing multiple pollutants from multiple sources simultaneously in particular places. Managing these complex interactions required more "administration" and the construction of custom-built, narrow "markets"—ones responsive to local hot spots, but by their nature unable to realize the efficiency gains of nationwide trading. These problems linger because their solutions demand more active regulatory collaboration, not less government. The market's eight good years show what trading can do, but also—especially in the case of severe environmental harms—what it cannot.

MAKING SCRUBBERS WORK

In theory, the logic of the Clean Air Act was simple. It mandated that the EPA set tough, science-driven NAAQS—standards that every locale would be required to meet, ideally prior to the nation's bicentennial celebration in 1976. Each state that had places in so-called nonattainment would be required to adopt a state implementation plan, which the EPA would review and approve.

Controlling pollution from processes deeply embedded in the modern industrial economy would prove to be a lot like today's problem of carbon dioxide, of course: extremely difficult and not amenable to ambitious solutions on a rapid schedule. Indeed, the dirtiest places didn't clear their air until well after 1976, and many are still in nonattainment today. But the Clean Air Act did gave the EPA a powerful weapon: it could reject a state implementation plan and dictate industrial policy in nonattaining areas. In reality, that weapon was politically and practically hard to wield, and wound up being reserved for egregious cases. Thus began a long negotiation that continues today between the EPA and the state regulators on which it must depend for the frontline work. While state regulators set local rules and checked progress, the EPA set goals and threatened penalties for lack of effort. In the words of William Ruckelshaus, the revered first head of the EPA who served as head again during part of the Reagan administration, federal weapons were essential to effective state action: "Unless they have a gorilla in the closet, they can't do the job. And the gorilla is [the] EPA."[37]

Scrubbers became an essential component of sulfur reduction only as other, apparently less costly control strategies proved infeasible. One strategy was to use less sulfurous coals. While all coal contains sulfur, its exact sulfur content varies widely—from less than 1 percent to about 10 percent. By far, the greatest concentrations of coal-fired US power plants were in the Midwest and South. Local coals there were generally high in sulfur. More distant coals were thought to have lower sulfur, but moving coal long distances in the 1970s was prohibitively expensive because of tightly controlled railroad tariffs. Local coals, especially at midwestern and eastern mines, also benefited from regulatory protection.

Beyond these factors, horrific mining practices—later pilloried as "mountaintop removal"—were tolerated at the time, and mining companies had few responsibilities to protect the local landscape. Nor were they required to post bonds for the restoration of sites when they were done. With no serious demand for low-sulfur western coal, there was no supply. The federal government owned vast mineral rights for western coals, but essentially none of that resource was leased until the 1980s, when the Reagan administration put politicians who favored the West in charge at the Department of Interior and a gold rush of western coal leasing began.[38] In the early 1970s, when minds were first focused on the problem of cutting sulfur from coal-fired power plants, the option of coal switching was not available at scale.

A second option was to clean the sulfur from the coal before combustion. That was tried and soon shown to be impractical as well. Also impractical at the time was to gasify the coal, which might have made it easier to remove gaseous sulfur before combustion; some demonstration programs dismissed that idea quickly. (Today, gasification is much better understood, thanks partly to a lot of innovation in coal-rich China, where gasification is a way to use coal as a feedstock for making chemicals. Coal gasification might be practical in the United States too, if the cost of natural gas had not plummeted when fracking and horizontal drilling unleashed massive new supplies, and utilities stopped building new coal plants.)

The impracticality of these two options meant that power plants and their utility owners had to face the reality that large amounts of SO_2 would be released from coal combustion. Compliance with NAAQS, which set limits on local concentrations of air pollutants, required breaking the link between sulfur released during combustion and what the local community experienced as sulfur pollution.

The third option was simply to build taller stacks, thus embodying the age-old maxim that the "solution to pollution is dilution." This approach would break the link between sulfur released from coal combustion and local concentrations of sulfur in the air, where it would cause harm to the local populations, by getting more of the sulfur to leave the immediate areas. Industry instantly calculated that taller stacks would lower local pollution, and they responded by retrofitting and building new, taller stacks. It was later learned that this practice actually worsened the problem of acid rain, which was the result of sulfur (and other pollutants) traveling days downwind and converting chemically into acids while aloft.

In the 1970s, acid rain was still a matter of scientific debate, not a top national priority, and unrelated to the immediate problem of compliance with NAAQS that were written entirely in terms of local air pollution. Nonetheless, environmentalists were alarmed by stack rising, which they saw as a ploy to avoid regulating pollution, and organized quickly to block the practice as an abdication of responsibility. In the 1977 amendments to the Clean Air Act, rules supported by the increasingly powerful environmental community barred taller stacks unless plants installed modern pollution control equipment too. Those amendments also included a scheme known as the "prevention of significant deterioration" that made sure that even the clean areas wouldn't get a commercial advantage by getting dirtier. It put industry on notice that new and existing power plants, even in cleaner parts of the country, would need to install pollution control equipment. The solution to pollution was no longer dilution.

What remained was a fourth option, and it became the most important for industry: pollution control equipment. For sulfur, that meant scrubbers. Coal plants are extremely complex machines, and their economic value comes from operating the machine as an integrated, reliable system. In reality, a plant is at least two systems that operate in an uneasy marriage. One is the boiler, which often takes hours to start and heat up. Inside the boiler, plant operators worry about temperatures and pressures along with the propagation of the flame across the fuel as it is injected. The boiler makes steam that is sent to a turbine, where another system functions; its operators have their own set of worries about the chemistry of the water, variation in temperature and the pressure of the steam, and the reliability of the turbine itself.

Each system, if it fails or gets out of synch with the rest, can shut down or throttle back the plant. Pollution control equipment adds one or more additional systems, more operations that must be synergized, and more things that if they operate unreliably, curtail plant output (and thus cost a lot of money). One of these is the electrostatic precipitator—a machine that sits in the flue gas after the boiler and shunts the gas over an array of electrified rods. The electricity charges particles, which stick to the rods. Periodically the rods shake, and the accumulated particulates drop to the ground (and hence don't go up the smokestack). After the electrostatic precipitator comes the scrubber.[39]

While everything in the front of the coal plant runs according to the principles of mechanical, combustion, and electrical engineering, the scrubber is a chemical engineering system. (Suspicions across these disciplines are legendary among operators of power plants. One plant designer, during a visit by one of us to a big coal plant, declared that "chemical engineering is the work of the devil.") The fear is that anything new bolted to an already-finicky engineering system will lower the reliability and raise the costs. In a scrubber, the sulfur in the flue gas reacts with water, reactive chemicals (e.g., limestone), and some catalysts to make prodigious quantities of some by-product (e.g., gypsum).[40] The engineering quickly demonstrated that all the components of a scrubber system would work. Bench-level lab work and small experiments generated a wealth of ideas about how to efficiently integrate scrubber components, but generating system knowledge in an industry where reliability is central required testing on actual power plants at scale, and collaborating with highly motivated owners that were willing to risk downtime and expense in exchange for knowledge.

The motivated owners were the firms with the biggest coal-fired power plants and most potential regulatory exposure, such as Detroit Edison, Southern Company, the Tennessee Valley Authority, Commonwealth Edison, and Southern California Edison. The early ideas for how to run scrubber chemistry came, in part, from the EPA's laboratories at Research Triangle Park, which did bench science and ran small pilots. Research Triangle Park has been historically critical to the EPA's pollution regulation because it gives the agency its own direct access to the latest science. The Tennessee Valley Authority—a government-owned corporation created during the New Deal to provide irrigation, electric power, and support for economic development in its region—also played an important role, including by running three ten-megawatt test plants at its Shawnee, Kentucky facility.

These test plants served as way stations where ideas that worked in pilots could be tested at modest scale before deployment at commercial coal plants sized in the hundreds of megawatts or larger. The Tennessee Valley Authority was also host to one of the first full-scale demonstrations, alongside Detroit

Edison (which carried out a pilot project in River Rouge, Michigan, followed by a full-scale demonstration in nearby Saint Claire) and several others. Around the mid-1970s, depending on how you count, there were between four and seven major contending scrubber strategies. Each demonstration was located at a plant with an owner motivated by a desire to understand and shape emerging regulations, and if successful, use the results from scrubber experiments to control pollution at a lower cost. Each utility feared that the failure to make serious efforts to understand and control pollution would wake the gorilla in the EPA's closet.

No utility would have been foolhardy enough to undertake this kind of experimentation in isolation from its regulators and peer companies. The risks that any single project could end in costly failure—while a similar one nearby in the industry succeeded—were simply too high. Active collaboration among utilities and with regulators was thus a precondition for exploration of the emerging technology.

Collaboration with the environmental regulator in particular was essential because plant operators wanted to show regulators what was in fact feasible (and hence should shape the standards). But collaboration was also important because the systems under test sometimes fail, causing more pollution. It is prudent to secure the regulator's forbearance for such well-intentioned accidents in advance—in return, for example, for the right to observe the process in operation. Over time, all the utilities learned that the consent of the environmental regulator would be extremely important for yet another reason: extremely costly plant upgrades could inadvertently be treated by regulators as creating a "new" plant—and under the Clean Air Act, depending on how regulators interpreted the law, new plants were controlled much more aggressively than old ones. The utility regulator was essential, too, because these experiments at scale on actual plants were expensive. With a regulator's blessing, they would be allowable research expenses (and therefore passed through to ratepayers without harming financial flows for stockholders) or, in some cases, investments (and thus capitalized into the rate base from which the utility and its investors earned a guaranteed rate of return).

Collaboration among utilities, meanwhile, was indispensable for all firms to learn the lessons of each plant experiment. Assuring that flow of information required engaging, beyond the firms themselves, two additional groups of actors. One was the cluster of equipment vendors and engineering contractors that could see across the totality of the industry—firms such as Peabody Process Engineering (which built the River Rouge pilot and then applied that knowledge to the full-scale Saint Claire plant) and Bechtel (which was the general contractor for many coal-fired power plants and also operated the Shawnee test facility). The other and perhaps more important group of actors was the industry itself, embodied in a newly created research collaborative: EPRI.

EPRI was itself the product of a penalty default. Fearing onerous federal regulation after the 1965 blackout, the electric industry set up a nonprofit collective research arm to develop its own solutions to collective problems.[41] The institute invited regulators and public interest stakeholders onto its advisory councils to help steer the research. Scrubbers were one of the first demonstrations of EPRI's essential role in collaborative research and information pooling. EPRI helped design demonstration projects and organize performance reviews by industry peers so that lessons from those projects about the cost and efficacy of different experimental scrubber systems became common knowledge in the industry; it also helped with "technology transfer" as firms absorbed the new knowledge into plant-level operations. (Using the same methods, the institute has played a central role in helping improve the performance of the US nuclear power industry.)

Through collaboration under the shadow of the penalty default, decentralized experimentation, and centralized learning through peer review, the scrubber was adopted as the best-available control technology for SO_2. Scrubbers in fact became a lot more reliable, and more acceptable to cautious plant operators and utility executives. New plants installed them. Even in extremely clean areas of the country, some plants installed scrubbers thanks to the prevention of significant deterioration (PSD) rule. Knowledge about scrubber reliability and efficiency led to refinement in regulatory rules; means and ends coevolved. In the early 1970s, installing a scrubber meant cutting sulfur pollution by 70 percent. With the 1977 Clean Air Act amendments, that number rose to 90 percent, and some local air pollution rules (under review by the EPA through state implementation plans) demanded even higher scrubber efficiencies—goals that were achievable at costs that would have been unthinkable prior to industry demonstrations.

It was not command-and-control decisions by remote bureaucrats that produced these results but rather an institutionalized process of collaborative experimentation and peer review involving firms, the industry's research institute, and state and federal regulators. In a word, the process was a case of experimentalist governance through and through.

Perhaps most striking is that the industry—organized through EPRI—needed knowledge about scrubber performance that didn't sit neatly within established ways of organizing knowledge inside the sector. Chemical engineering mattered, of course, but the best scrubber systems turned out to be those that involved reacting sulfur with limestone—both because limestone was cheap and available almost everywhere, and because good chemical control could convert the product of limestone-sulfur reactions into gypsum, which was valuable (as wallboard) or easily discarded. Through the early 1980s, alternative systems based on chemical reactions with Mag Lime (containing large amounts of magnesium oxide) were thought to be viable, but they faltered on the lack of cost-effective supplies and the expense of waste removal. Only by the mid-1980s was the limestone tower sprayer scrubber system widely accepted as best under almost all conditions.

Ironically, this collaborative approach to technology forcing was well established by the 1980s. Scrubbers were its most iconic success, and with their introduction, sulfur emissions were already declining quickly. From 1970 to the late 1980s, emissions had dropped by about ten million tons annually. But the successful continuation of collaboration and translation of the results to standards required the willingness of the EPA to continue to invoke the threat of a penalty default, and Congress to keep the Clean Air Act framework in place. The rising tide of deregulation ultimately made constancy impossible. In the first two years of the Reagan administration, EPA head Ann Gorsuch tried to eviscerate the agency she headed. When Ruckelshaus took over again in 1983, normalcy returned and collaboration continued. Yet the interlude presaged what was to come for much of the George W. Bush administration and all of the Trump administration.

THE TURN TO MARKETS: NEW ENVIRONMENTAL CHALLENGES AND NEW POLITICAL FORCES

In the late 1980s, the political stars aligned for another big change in federal law on air pollution—what became the 1990 amendments to the Clean Air Act, adopted with overwhelming bipartisan support. This amendment, the first since 1977 and the last Congress has ever passed, allowed for the codification of new missions—in particular, efforts to control acid rain. And it allowed for the act to be updated to include a new suite of policy approaches: pollution markets.[42]

From the 1970s on, the problem of acid rain was known in theory, and evidence was accumulating that fragile ecosystems, especially along the Eastern Seaboard downwind of industrial pollution, were deteriorating. Similar evidence appeared in Europe as well, made visible by *Waldsterben*, the dieback of forests concentrated in central Europe that accelerated in the late 1970s and helped galvanize public opinion in favor of pollution control. Many hypotheses were offered for these dramatic declines in ecological health, but the exact causes were unknown. In 1980, the last full year of the Carter administration, Congress created the National Acid Precipitation Assessment Program to study the problem over a decade, and report on the impacts, costs, and solutions.[43]

Meanwhile, political momentum for action grew and crested throughout the Reagan administration. Unrelated to air pollution policy, during the 1980s, the government also instituted reforms in coal leasing (which opened bigger supplies of low-sulfur coal in the West) and implemented a 1980 railroad deregulation law that lowered rail rates by 60 percent over the decade (which made it much cheaper to ship low-sulfur coal from the West to power plants around the country).[44] Had the country continued to deploy scrubbers and invest in better scrubber innovation, it is plausible that sulfur emissions would have come down just as quickly, if not faster, than required by the new limits inspired by worries about acid rain. But the politics of pollution control were shifting, and building a bipartisan consensus required a shift to markets.

The idea of using markets to control pollution was not new. The theory dated back at least to the 1960s, and in the 1980s, a coalition of academics, environmentalists, and industrialists helped articulate to policy makers the benefits and modalities of market-based pollution control.[45] A large lead-trading program had demonstrated the practicality of cutting compliance costs through market strategies—albeit in the context of a rapid phaseout of a pollutant for which substitutes were already well-known.[46] Under the 1977 amendments of the Clean Air Act, there were limited opportunities for the trading of pollution offsets and credits.[47] Despite the limits of this experience, retrospective studies showed that markets usually saved money.[48] In 1987, there was a major legislative push to amend the act to address acid rain by requiring more extensive use of scrubbers. That failed, in part, because technology mandates were seen as costly intrusions by government.[49] At about the same time, there was bipartisan support in the late 1980s for the greater use of markets for achieving environmental goals.[50] For those who were skeptical of government effectiveness and yet still interested in pollution control, markets were the way forward.[51]

The market solution relied on a cap-and-trade system. At the time, US emissions from large power plants were about twenty million tons of SO_2 per year.[52] The National Acid Precipitation Assessment Program had not reported its results yet, so it was impractical to do a cost-benefit analysis.[53] The best, highly uncertain estimates suggested that the costs of emissions reductions would rise steeply as abatement exceeded ten million tons (a cut of power plant sulfur pollution by about half). The cap would apply starting in 1995, and big polluters that acted early could earn valuable credits that they could bank for future phases of the program when it was expected that the costs would be higher. To avoid creating too many enemies, emission permits were given away mostly for free to existing polluters.[54] Because firms knew that grandfathering was going to be the likely rule, in the late 1980s, when elements for a national trading system were being debated, the continuous decline in sulfur emissions actually stopped. Polluters saw more pollution as an advantage, and made no additional efforts to cut emissions until the baseline (1990) was agreed and an economic advantage from further cuts was written into law.

Analysts of cap-and-trade systems have looked at details of the sulfur market design with an eye to how they create economic efficiencies. Early action, which would lower emissions promptly and create extra credits that could be banked for a later phase, allows for greater efficiency in the timing of emission controls. Detailed monitoring requirements—through the continuous emission monitoring systems equipment that all large polluters were required to install—would make sure that behavior was transparent. Avoiding new requirements for equipment mandates would allow for technological efficiency. Putting the whole country under a single bubble would maximize the gains from trade. The free allocation of permits to emit, if necessary to avoid political blocking by utilities and coal suppliers, was acceptable because it would not harm economic efficiency; the giveaway just affected the allocation of policy costs in ways, hardly surprising in politics, that benefited incumbents.[55]

ACHIEVING STATIC EFFICIENCY

For about eight years, the national sulfur market worked. Prices were relatively stable, and the total costs were less than half of the original expectations.[56] In the conventional view, markets created this tremendous success.

But assigning causality is, as always, tricky. Broadly speaking, the sulfur market allowed generators to choose among three options to comply with caps on emissions: buy credits on the marketplace (or equivalently, use their own credits that had been banked), install and operate scrubbers, or switch coals. Two out of those three options were made possible by actions outside the cap-and-trade program.[57] Indeed, the most important innovations in sulfur removal—scrubbing—predated the cap-and-trade system; it was the product of experimentalist governance. And while advocates for market-based policies often claim that these approaches encourage technological innovation, careful analysis of this history has shown that regulatory approaches forced a lot more innovation (and patenting) compared with more market-driven approaches.[58]

Because firms were free to make these choices, the market helped achieve some static efficiencies. In political terms, however, these choices actually created substantial harm because they made learning and adjustment politically harder to achieve. By fixing pollution caps at the outset, they locked in the expectation that pollution control costs would be high. When participants learned that compliance costs would be much lower than projected, it was impossible to adjust—to make the actual effort align better with what society was willing to pay for pollution control—without changing the statute. The political preconditions for an efficient market—high credibility with clear limits on pollution—interfered with its efficient operation.

But these questions aside, even as the cap-and-trade system entered its salad days, another fundamental limit on the utility of markets was coming to light.

THE LIMITS TO MARKET-BASED APPROACHES TO SOLVING COMPLEX, LOCAL POLLUTION PROBLEMS

The general problem was that national sulfur controls—creating wide and efficient markets—were mismatched with the real environmental problems, which often involved local mixtures of many pollutants. Concretely, acid rain was the result not just of SO_2 but also its combination in particular air sheds with NOx that turned to acid. Solving the acid rain problem would thus require addressing NOx as well. And since coal-fired power plants were not the only sources of those pollutants (half of such oxides came from vehicles, for example), any effective strategy would need to address those other sources of pollution in addition to large stationary coal-fired power plants. Moreover, just as the 1990 Clean Air Act amendments were being adopted, it became clear that there were many other regional air pollution problems, such as smog and haze.[59] Over the years, still more regional pollution problems would become understood too, such as emitted mercury. These were the real problems that needed solving; the sulfur market, by design, focused on something different: nationally averaged emissions.

To be fair, this mismatch was discussed—if not fully appreciated—when the national sulfur program was created. Ideas were floated, for example, to address SO_2 and NOx together in a single market, but they were rejected as too complex. Complexity, it was rightly thought, would undermine the ability of the market to send simple and clean price signals. During the period of the most active reduction—the golden age of the sulfur market—the fictions needed to make the market work were not too debilitating. The total pollution was tumbling. And the counterfactual for comparison—the speed at which emissions would have declined if the experimentalist system of collaboration had just stayed in place or had been perhaps expanded, as contemplated in 1987, with requirements to use scrubbers more widely—was unobservable.

But once the market had helped achieve quick reductions in pollution, the disconnect became readily apparent. Subsequent attempts to revise the market solution revealed that when pollution problems are highly complex and poorly understood, it is impractical to create a credible market design that can respond to the problem at hand. There were many attempts to do just that; all failed, and market participants soon learned that *faute de mieux*, the old methods of the Clean Air Act—regulation—remained as the key tool for controlling pollution. The sulfur market was no longer credible, and prices tumbled to zero.[60]

One cause of the failure, already feared by many environmentalists when the nationwide market was created, was that trading based on national standards created hot spots. Trading might lower costs for abating sulfur overall nationwide, but by design the effort would differ across plants. In places where pollution control was relatively inexpensive (e.g., at plants located closer to cheap, western, low-sulfur coal or plants of more recent design that had scrubbers installed), the emissions would tumble. The emission credits thus generated would be sold to dirty places and used to keep emissions high or let them go higher.

One retrospective study has shown this fear was justified: western, newer plants did cut more than older plants, which tended to be located near cities. The cost of those higher local sulfur emissions (where they caused harms to human health) were five times the static efficiency gains from the sulfur trading system overall.[61] Looking back, it seems that even by the standards of a single national goal—the easiest way to create a national sulfur market—the sulfur trading program was less efficient and possibly value destroying compared with a plausible experimentalist alternative that would have made the best efforts to apply scrubbers. Alas, much of this story has been pieced together only recently, in part because of an improved ability to measure local health effects from sulfur and other pollutants.

The other cause of failure is the one that actually undid the sulfur trading program: the complex interaction between sulfur and other pollutants, including NOx, in the creation of regional pollution. The level of control needed to vary with the level of downwind pollution and ecological sensitivity. This problem arose during the Bush administration, which embraced markets on ideological grounds, crafting a complex scheme that would allow for the trading of multiple pollutants with varied weights depending on geography. Creating this kind of a market would require a lot more information than a simple national cap, and much of that information, such as the exchange rates between pollutants, was either unknowable with precision or varied in such complex ways that it would be impractical to design a market aligned with the exact, changing nature of emissions and pollution impacts. Nonetheless, the Bush administration made a formal legislative proposal to do just that: the Clear Skies Initiative. When that initiative failed politically, the administration attempted to achieve the same outcome through administrative action in the form of the Clean Air Interstate Rule.

Ideologically committed to using markets, the Bush team created ever more complex contortions that never could align incentives with the real nature of the environmental problem and control at hand. But so strong was the impetus to use markets that prior experiences to do exactly this—experiences that showed the difficulty and meager rewards from such markets—were ignored. For example, in the 1980s, a team of economists had been asked to design a market to control noise pollution at airports. Tom Schelling, who led the effort, reported that the information needed to make the market work—the exact properties of different aircraft types, flight paths, sensitivity to noise, and so on—was essentially identical to that needed for direct regulation that would not use market forces.[62]

This was the beginning of the end of the sulfur market. After pivoting away from Clear Skies to the Clean Air Interstate Rule, the process of review and sources of penalty defaults also shifted to the courts, which threw out the Bush market initiatives because they strayed too far from the original purpose of the Clean Air Act: to cut pollution, reliably, anywhere and everywhere needed. Bush's EPA never finished its attempt to update its failed rule. The incoming Obama team, working through administrative action as well, offered its own proposals, but they too failed to deliver what the courts and political leaders saw as essential in protecting the environment. As Clear

Skies, the Clean Air Interstate Rule, and the Obama proposals evolved and died, market participants quickly learned that the old sulfur market was no longer credible. The golden era of sulfur trading was over.

Less visible and more important was that the Clean Air Act's original mission was still in place, and the EPA's gorilla would be awakened periodically. Industry knew this, and it kept collaborating as it had on sulfur—working on each new frontier in pollution control. For instance, the same lead firm (Southern Company) working with the same industry organization (EPRI) tested multiple methods for cutting mercury.[63] The logic remained the same: better to get ahead of the gorilla and collaborate than wait for it to come crashing in the front door.

What are we to make of this story? Some market advocates see the collapse of sulfur trading as an object lesson in government's fickle nature, sometimes giving (creating a market) and sometimes taking it away. We see this history differently. The market can thrive so long as there are many opportunities for trading—a large trading zone with a single price that homogeneously affects all actors—and the participants are choosing among options with a known performance. Yet the real challenges and goals in controlling pollution are much more complex, as are the technological and operational changes needed to clear the air.

It would be a misreading of sulfur history to conclude that either markets or administration are inherently ineffective. The real lessons are about where and why different policy strategies work best. When uncertainties about control strategies are large and the best strategies forward are unknown, experimentalist regulatory systems work best. All parties have an incentive to decompose the problem into smaller, solvable components and work on solutions; the regulator knows that goals must be treated as provisional and adjustable in light of experience. Once goals can be set more firmly, and the range of technological and behavioral options is better understood, markets can take over. (Under those conditions, however, much of the success of markets is also available to regulators because the best choices are widely known.) That happened in the history of sulfur control, but recalling only that portion of the story distorts the moral, suggesting that regulation was the problem and markets were the solution. As we've shown, a fuller account demonstrates the crucial role of experimentalist regulation in making basic innovation practically applicable and revealing the limits of markets in solving local problems. Cap and trade was useful in the interlude when it worked, but as a promise of fundamental reform—a driver of innovation—it failed.

Innovating at the Border of Science and Technology

We turn now to our final case study of ARPA-E. Much of the time, technology-forcing regulation implemented through experimentalist processes can induce or accelerate the innovations needed to address policy goals. But not always. When innovation is especially costly and risky, the payback period for successfully commercialized products is especially long and uncertain, and the full benefits of innovation are hard to appropriate, then the motivations for innovators may be insufficient. In the case of such market failures, it is the role of government to step in and directly encourage as well as monitor investment in those key technologies and capacities without which progress on a broad front will be impossible. In the vocabulary of chapter 3, the form of intervention switches from regulation to industrial policy.

The problem, perhaps especially in the US context, is that alongside the arguments in favor of industrial policy as a response to market failure, there are arguments about the likelihood of *government* failure that make such industrial action fraught with danger. All the reservations about the shortcomings of administration and all the cautions about regulation, and particularly regulatory capture, count greatly with respect to industrial policy. Public bureaucracy is too sclerotic, the complaint goes, and the political process is too beholden to special interests to allow for dispassionate, expert decision-making, even supposing that public officials can know enough about technology and markets to have the relevant expertise in the first place.

Sure enough, industrial policy is one of the most contentious topics in economic policy. Opponents insist that arrogant ambitions to "pick winners" will produce "white elephants" or "cathedrals in the desert." Proponents respond that in the face of manifest and severe market failures, the costs of inaction far outweigh those of this or that bad decision. The argument is typically by example—with failed investments such as Fisker (an early electric vehicle start-up) and Solyndra (a company searching for solar cells that would require less silicon) thrown into one pan of the balance, and successes such as Tesla (which repaid a loan nine years early) stacked on the other. Similarly, every study decrying a "technology pork barrel" can be paired with another portraying the (sometimes) proven promise of industrial policy.[64] Even in development economics—which coined the term "industrial policy" to connote the bundle of measures by which a poor economy could build the industries that were long thought to be the precondition for sustained growth—there is surprisingly little attention given to the governance arrangements that explain why these policies do and don't work.[65] Yet it is these governance arrangements that hold the key to understanding when and how industrial policy gets the job done.

One essential element of these arrangements is their impact on technological innovation. Successful industrial policy must be anchored in a strategy for supplying fundamentally new ideas. (The case of climate change is particularly emblematic; a recent study by the International Energy Agency concludes that about three-quarters of all the technologies needed for deep decarbonization aren't yet technologically or commercially viable.)[66] ARPA-E, which began operation in 2009, provides an illustration of a highly successful system of clean energy innovation.[67]

ARPA-E's overarching goal is to eliminate "white spaces" in the map of technical knowledge: missing capabilities just beyond the frontier of current technical possibility that if mastered, would clear the way to advances in an important domain. A program might, for instance, aim to support the investigation of novel battery concepts with the potential to reduce storage costs enough to make an attractive class of electric grid designs economically feasible. The agency fills an important hole in the system of energy innovation supports—between publicly funded start-ups, which emerge from universities or national labs, and privately funded ones, which emerge through venture capital and incubators.[68] One indication of ARPA-E's robust success is that despite regular political accusations that it is wasteful and crowds out the private sector, it can count for funding support on a bipartisan group of legislators impressed by the results. For our purposes, ARPA-E is crucial not only because of its role in clean energy but also because its systems for governance have been studied closely.

ARPA-E inherited and then refined its governance structure from its near namesake, the Defense Advanced Projects Agency (DARPA), created in 1958 in response to the Soviet launch of the Sputnik satellite. In discussions of industrial policy, DARPA is often invoked as a reminder that the state played, and continues to play, a fundamental role in organizing the research from which are hewn the building blocks of the information economy. Among its iconic contributions are the computer network protocols underlying the internet, precursors to global positioning systems, and fundamental tools and devices for microprocessor design and fabrication. The accomplishments of DARPA have inspired a number of research agencies along similar lines, of which ARPA-E is both the most successful and most faithful to the procedures of the original model.[69] (Of course, there are significant differences between DARPA and these descendants. One is that DARPA is oriented around more extensive discussions with customers and users of technology, and about half the DARPA budget is spent on later-stage testing and evaluation along with the early deployment of technologies. ARPA-E, by contrast, focuses almost entirely on early stage innovation, in part because there are other programs at the DOE that work on the later stages.)[70]

At every stage in the organization of research—defining the target or programs of investigation; selecting projects that advance the program purpose; and supervising individual projects in the program portfolio—ARPA-E treats goals as provisional or corrigible, and uses peer review to surface differences triggering further investigation. To begin with, program directors are hired largely on the basis of their promise in giving shape and direction to an emergent area of investigation. A candidate with a background in geology, for example, will be hired to create and concretize a program in advanced geothermal energy. Once the program goals have been framed, the program director does a "deep dive," supplementing and correcting their own background experience with reviews of the scientific literature, site visits to universities and companies by ARPA-E technical staff, commissioned external studies, and consultation with research managers elsewhere in the DOE. Program directors then test the practicality of the emerging research area in technical workshops involving leading engineering, scientific, and commercial experts. If the research plan, adjusted to reflect the exchanges at the workshop, passes review, a project is formally created as a component of the developing program.[71]

Proposals for research within the projects are developed and executed in the same manner, with the goals open to recurrent challenge and revision via peer review with ARPA-E's professional staff. Applicants first submit a concept paper: a short document explaining why the proposal is superior to alternative approaches, and how it responds to foreseeable technical and commercial risks. Proposals that survive a first round of external review are developed into full applications and reviewed again, with the difference that applicants may rebut criticism by external reviewers.[72] The winners, designated as "research partners" or "performers," then negotiate project milestones with the agency staff.

A notable and perhaps surprising feature of this process is that the selection is not based on a consensus view of the project's prospects. Holding the rating constant, the agency picks the project where the range of reviewer rankings is the *greatest*—that is, where judgments diverge the most. Since disagreement among experts is a good working definition of uncertainty, this preference for divergent rankings is a further indication that ARPA-E fully recognizes the uncertainty in its environment. In the absence of expert consensus, the managers and selection committee rely on other information, such as rebuttals, observation of the research in workshop, dialogue with peers, and the need for methodological diversity when the best strategies are unknown.[73] Managers know that consensus-seeking peer review—common in academia—often has a bias against the novelty and risk that transformative research programs deliberately seek.

Once funding is offered, ARPA-E managers remain intensely involved—checking milestones, adjusting plans, and learning. Missed milestones can touch off an intensification of site visits, conference calls, meetings, and written analyses of problems and possible solutions. When projects struggle, milestones can be reset to permit an alternative to the failed approach. Milestones are added or deleted in fully 45 percent of the projects, not counting substantive modifications, which are said to be frequent. Staff can adjust the length and budget for projects (empirical work finds that most projects see a modest lengthening and rise in budget). If recovery efforts fail, the program director sends an "at-risk" letter warning of the possibility of termination.[74] In short, the agency rejects the model of hands-off, bet-on-the-person-not-the-project administration preferred by many established and successful research funders, public and private, in favor of the continuous, collaborative review and adjustment central to experimentalist governance, and widespread, as noted in chapter 3, in biotechnology, advanced manufacturing, and venture capital.[75]

ARPA-E is a particularly successful model of experimentalist governance in an organization (the DOE) and domain (energy research) often seen as anathema to nimble experimentalism. Yet ARPA-E is not alone. Within the DOE, there are many other technology demonstration programs, and they too have learned how to elicit ideas from the outside, define progress in terms of commercial potential, and adjust those terms in response to new information. The results from these investments suggest that the pork barrel critique, if it was ever valid, is outdated. A large statistical study of direct DOE support to early technologies through the small business innovation research program has found that the program does increase the prospects for commercial success; comparing proposals on both sides of the line between the DOE's support and nonsupport, the ideas that get the halo and resources of the DOE backing do better.[76] (Other government agencies run similar programs, with encouraging results.)

A separate, independent study of US cleantech start-ups has shown that those with government partnership fare better, in part because of quality signals from the partnership; the patenting nearly doubles and private financing rises 155 percent for start-ups that have government alliances compared with those that don't.[77] From these numerous, diverse, and encouraging experiences, a whole set of "best practices" has emerged about how to use government research, design, and development resources effectively—a role for intensive peer review, centralized periodic evaluation of portfolios, revision of goals in light of experience, attention to commercialization metrics that are adjustable with experience, and a blend of decentralized effort but centralized assessment.[78] Those metrics are, in a different word, key elements of experimentalism.

Organizing Innovation

What are we to take away from these three case studies? CARB is dedicated to technology-forcing regulation, the sulfur trading regime to market making, and ARPA-E to a central element of industrial policy: innovation. Together these three strategies constitute the principal choices for climate change intervention, and often these choices are subject to bitter political contest, on the assumption that selection of the "right" strategy for a particular context was necessary and sufficient to ensure success.

Our argument is that these contests are fighting the wrong fight and arguing over the wrong question. There is no knockdown argument to be given for the "right" choice of intervention—at least if our goal is practical success. Settling on one of these strategies is insufficient because it leaves unspecified the all-important governance arrangements that determine whether decisions can be rapidly corrected under uncertainty. In the end, what made CARB, ARPA-E, and sulfur control successful—setting them apart from efforts otherwise similar in their respective strategies—were their structures of experimentalist governance. Experimentalist governance, in short, is a necessary condition of success (though hardly sufficient).

At times these diverse policy strategies may be complementary. We saw in the case of sulfur that regulatory systems that pushed technological innovation were complementary to market incentives. We suspect that this is frequently and perhaps regularly the case; regulations and active investments in innovation are the anchors of industrial policy, and market incentives help optimize the system once directions and technologies are known.

Equally important to the technical processes of regulation under uncertainty are the political effects of these programs. Both CARB and the sulfur program explicitly evolved in ways that kept solving environmental problems at an acceptable cost, which helped to sustain political support. The only big exception to that rule was the brief period when nearly all sulfur reductions occurred through the cap-and-trade program; then the actual effort fell short of what the public probably would have tolerated and drifted away from what was needed for environmental protection in key jurisdictions, such as the states where power plants were located close to vulnerable populations. This political logic to experimentalism is evident as well at ARPA-E, where some of the new technologies emerging from ARPA-E support are offered as evidence that transformations in energy markets can run faster than was originally thought with previous generations of technology.

The cases discussed in this chapter all share a feature in common: they are examples of innovation at the technological frontier. In fact, a great deal of innovation takes place apparently far from the frontier when new products or processes are often adjusted—and sometimes reinvented—to work under local conditions for which they have been only partially designed. Such localized innovation, or contextualization, is particularly important in climate change, where solutions and institutions designed for one place regularly don't work well in others. We turn to this contextualized innovation in the next chapter.

5

Experimentalism in Context

GROUND-LEVEL INNOVATION IN AGRICULTURE, FORESTRY, AND ELECTRIC POWER

In climate change and decarbonization—as in education, health care, and the provision of electric power—the problems are general, but the solutions are often local. What works in one place doesn't work the same way or at all in another. Innovations at the frontier have to be adapted and sometimes transformed to suit the peculiarities of place. And because place-specific conditions are constantly changing, local adaptation rarely comes to an end. We call this ongoing adaptation of innovations to the economic, social, and political circumstances of a particular place *contextualization*.

The process of local adaptation is especially conspicuous and intrusive in climate change. The effectiveness of pollution mitigation measures varies with the seasons and climate as well as with new ways of working and living. At the same time, these measures intrude deeply into everyday life, changing how property is used, people get about, industrial products are made, or how much water is available. In climate change, think of contextualization as the way decarbonization takes root in the endlessly varied soils of everyday life.

This chapter presents three case studies that exemplify the kind of contextualization that will be necessary as industrial and agricultural systems are transformed along the lines needed for deep decarbonization. First, we consider Ireland's contextualization of EU water pollution standards through a dynamic framework of local governance. Second, we examine California's contextualization of renewable energy, highlighting the collaboration between state regulators and local utilities and the deployers of new technologies that are transforming the state's grid. Last, we discuss Brazil's efforts to contextualize the idea of sustainability by devising promising ways of reconciling growth and the protection of the environment in the Amazon.

These examples of contextualization run contrary to conventional thinking about the local deployment of laws and technologies. In standard accounts, the design or conceptualization of a new machine or policy is radically separated from the process by which it is installed or implemented locally. The aim is the local replication of a model or realization of a blueprint, and success is measured by the fidelity of the new instance to the original. These accounts portray the actual process of implementation as ineffable, drawing on tacit, local knowledge instead of explicit practices of deliberation. On this picture, as in the assembly of a complicated piece of furniture, instructions point much of the way, even if they must be interpreted in light of a user's unspoken assumptions about what the diagrams mean to show, or how to work with the relevant tools and materials. Frustration is inevitable and idiosyncratic.[1]

In other words, in this standard account, implementation is ultimately a matter of brute improvisation; there is virtually nothing systematic to be said about it. Even studies that explore how developing countries implement technology from more advanced nations treat the struggle to configure equipment to local conditions as a daring uncertainty—a venture frequently undertaken only out of ignorance of the pitfalls ahead, and successful only because adversity inspires ingenuity.[2] Likewise, efforts to implement legal frameworks are conventionally seen as raw struggles for power—an extension of the political fights that led to the measure in the first place. The struggle may shift the outcome closer to the original preferences of one side or the other, but it does little or nothing to change views of a good solution.[3]

This view of the spread of new solutions has been repeatedly challenged empirically. From an experimentalist perspective, the premise that we can separate the conception and execution of large and complex projects is theoretically dubious. As the case studies in this chapter show, innovation spreads not by the installation of finished products or transposition of fixed legal provisions but rather through the continuation of the initial process by which technologies and legal measures were designed. Installation and implementation shade into reinvention. The tacit choices about adaptation implicit in implementation become explicit as they are subjected to continuing, deliberate discussion. Under the current conditions in climate change, this shift can have implications for the process of local decision-making as well. As traditional regulation strains and sometimes breaks under the pressure of continual local revision, rules can be adjusted and reinterpreted. Ad hoc administrative adjustments thus give rise to changes in governance in the direction of experimentalism.

The heart of the contextualization efforts we recount here lies in a special kind of collaboration: joining experts of various kinds and ranges of authority with each other as well as with local community members in the kind of reasoned give-and-take more typically associated with the practice of science. These forms of inquiry are conducted outside the laboratory and therefore are sometimes called "cognition in the wild."[4] But they are as disciplined as science in the use of peer review and constructive criticism. Deliberation in the wild does not eliminate struggles for power, any more than the practice of science fully banishes it from the lab. But the case studies show that when conventional alternatives have failed and penalty defaults loom, the availability of a deliberative framework can transform raw disputation into meaningful local learning.

The European Water Quality Directive and Irish Agriculture

Our first illustration of effective contextualization is Ireland's management of agricultural pollution over the last two decades. Within environmental regulation, nonpoint source pollution provides the best example of the uncertainty that results when familiar conditions, each well understood in isolation, combine in unforeseen ways. Emissions from large polluters, such as power plants or sewage treatment facilities, are relatively easy to detect and control. Much more problematic are intermittent emissions from diffuse sources, such as the runoff from sporadic detergent use in scattered households. Agricultural runoff from manure, excess fertilizer, and pesticides is especially refractory because conditions vary widely among multiple sources and along the paths of pollution.

Today, Ireland is at the forefront of innovation in the control of agricultural runoff. The highly competitive dairy industry is a major polluter of water and, because of ruminants' digestive gases, air. Ireland cannot meet its emissions targets—nor can the industry expand to realize its potential—unless this pollution is controlled. Though industry and governance have often been at loggerheads, at present they are highly motivated to find mutually workable solutions. Ireland's innovative system for the collaborative, local governance of environmental problems is the product of these pressures and opportunities.

The conviction that dairying could be a modern engine of growth came late to Ireland. Through much of the twentieth century, Irish dairy farming, like Irish farming generally, was dominated by extremely small holdings, with limited export opportunities along with relatively low productivity and incomes. When Ireland joined the European Economic Community (the predecessor of the European Union) and its Common Agricultural Policy in 1973, membership expanded market access and raised prices, leading to increased output and productivity. The imposition of EU milk quotas in 1984 prompted consolidation, yielding fewer but more efficient and specialized dairy farms that were still small—measured by farm acreage and herd size—in comparison to industrial producers. Irish dairy cooperatives, which process the farmers' milk, consolidated in this period and grew rapidly to become first-tier suppliers of ingredients to global consumer food firms, many of which built their own processing plants in Ireland as well.[5] In 2017, the country—which accounts for less than 1 percent of the global milk output—supplied almost 10 percent of the world's infant formula market and was the second-biggest exporter of infant formula to China.[6] Altogether, Ireland exports 90 percent of its dairy output.[7]

Many factors have contributed to the success of Irish dairying. First, Ireland enjoys a significant competitive advantage thanks to its natural supply of grass. A typical large Irish dairy farm has the lowest cash cost-to-output ratio of the key international milk-producing regions, including the United States, New Zealand, and Australia.[8] The key is homegrown grass feed, which is cheaper than purchased feed. The price of grass feed is also more stable than the price of purchased feed; relying on local grass thus shelters Irish dairy farmers against a substantial risk. Second, cows that pasture on grass, instead of consuming feed based on grains and soy, produce milk solids of superior quality.[9] Indeed, the grazing cow is the emblem of food production at its most natural. Third, dairy farming requires proportionally fewer imports than the transnational pharmaceutical and information technology firms that dominated the Celtic Tiger boom before playing a large role in the country's downturn during the Great Recession of 2008. Moreover, the profits of domestically owned dairy firms remain in Ireland, while those of high-tech firms are repatriated to foreign owners. Per unit of output and exports, the agri-food sector thus makes a larger contribution than the sectors dominated by foreign investment to the balance of payments and employment as well to regional and rural development.[10]

For all of these reasons, both the Irish dairy sector and its counterparts in various government departments have come to embrace the national system of grass-based dairying on family farms. Dairying has therefore found a central role in the overall development of the country—provided it can reconcile increasing efficiency with regulatory and consumer demands for environmental sustainability.[11] Wariness of or outright resistance to climate change mitigation is giving way to active collaboration in developing new measures and institutions.

The high-level regulatory frameworks governing these efforts go back decades. The European Union's Nitrate Directive of 1991 was one of the first measures to protect water quality from pollution by agricultural sources. Highly prescriptive, it set out precise nitrate concentration limits transcribed in each member state's Nitrates Action Program. Farms that fail to comply can be fined or disqualified from the valuable EU single farm subsidy. Countries that fail to meet national limits must submit a plan for improvement to secure a temporary derogation of requirements or face the potential application of draconian sanctions typical of penalty defaults.

The Nitrate Directive with its concentration limits became an integral part of the encompassing 2000 Water Framework Directive (WFD), which is, however generally, much less prescriptive and more experimentalist in character. The broad objectives of the WFD are "good water" (including minimal pollution by the listed chemicals) and "good ecological status" (where the target status for each type of water body—such as alpine streams or freshwater lakes—is minimal deviation from an ecological norm associated with a pristine water body of that type). An "intercalibration" procedure assures that countries apply comparable standards.[12] The basic unit of management is the river basin or catchment: the contiguous territory that drains into the sea at a single river mouth, estuary, or delta. Member states produce a six-year river basin management plan for each basin by a collaborative process in which public officials, experts, and stakeholders specify objectives as well as procedures for translating them into concrete activities. Each member state appoints a water director to oversee the execution of the plans. Together the water directors form a council that in consultation with the European Union's executive body, the European Commission, directs the preparation of guidance—known as the Common Implementation Strategy—in the application of the directive. Until 2027, countries that fall short can submit a new river basin management plan at the end of each planning cycle on the grounds that the earlier approach proved technically infeasible, disproportionately expensive, or was obstructed by extraordinary natural conditions. Thereafter, as a penalty default, cost and feasibility will not excuse noncompliance.[13] In short, central targets and corrigible penalties, though severe, are contingent on effort.

The implementation of both the prescriptive Nitrate Directive and more experimentalist WFD has proved frustratingly difficult. The adherence to "good practices" in agriculture, for instance, has often failed to produce improvements in nitrate levels; the effective, inclusive participation of local actors in the definition and continuing revision of the intentionally open-ended goals has been a major stumbling block in the application of the WFD. The Common Implementation Strategy has been revised many times.[14]

In Ireland, in particular, these kinds of failures triggered a series of research programs under the directives aimed at deepening the understanding and control of pollution flows at the catchment and field levels. These programs—buttressed by the findings of similar ones in other member states—have in turn helped generate a web of experimentalist institutions. The result is an integrated system for the local governance of water quality, greatly expanding public participation in environmental decision-making in the process.

The first and most important of these investigations was carried out under the Agricultural Catchments Programme, established in 2008 by Teagasc, the Irish agricultural research and extension service, in preparation for Ireland's application for the derogation of some requirements of the Nitrate Directive. Six catchment areas, differing in soil types, geology, and types of farming, were selected to monitor and model the relations among farm management practices, the migration of nutrients from their source to various water receptors, and the resulting changes in water quality. Some three hundred farmers participated in the program, each supported in the development of a nutrient management plan by a Teagasc extension agent, who could in turn draw on the additional expertise of fifteen researchers dedicated to the project.

The Agricultural Catchments Programme found, surprisingly, that local variation in the absorption and drainage of nutrients renders general rules ineffective. Poorly drained fields with phosphorus values too low for cultivation may still pollute because of fast surface runoff, for example. Conversely, soils with phosphorus concentrations in excess of agricultural needs, and therefore likely to burden the environment, may not pollute at all because they are especially well drained.[15] The policy implication is that a nutrient management plan should be only a starting point or provisional guide for an investigation in which a farmer and adviser collaborate in identifying the problems of a particular farm, devising remedies, and jointly monitoring the results.[16] A second catchment study, undertaken by the Irish Environment Protection Agency (EPA), confirmed the Agricultural Catchments Programme's findings, extended the investigation of the mechanisms of pollution transport to the geological structures beneath the soil layers, and showed that the disruption of pollution pathways is frequently a more effective means of mitigation than attempting to eliminate pollution at its source or to contain its effects at the receptor.

The new catchment program is part of a larger effort by the Irish EPA along with its partner institutions in water quality management to establish a cascading process of national, regional, and local consultations. The goal of this network is to select priority areas for intervention, and create local governance institutions to support the execution of the agreed-on interventions with the full and effective participation of the affected actors. The selection process and new governance institutions come together when the priority areas are subjected to a final, searching review in "local catchment assessments": field-level examinations by the local actors themselves of the sources of pollution in given water bodies. This assessment determines the local work plan, specifying, costing, and prioritizing projects. Such a collaborative review of priorities is particularly important in rural areas. In such regions, many small and frequently diffuse sources of pollution can confound mitigation, and deep local knowledge is indispensable to a deliberate and consensual choice of which problems to attack, and in what order.[17]

These efforts were formalized in 2016 with the establishment, as part of Ireland's transposition of the requirements of the WFD into national law, of the Local Authority Water Programme (LAWPRO) as a shared service between all local authorities in cooperation with the Irish EPA and other state bodies. LAWPRO supports catchment assessment and pollution mitigation by providing technical assistance to the stakeholders and helping each locale achieve inclusive engagement in the implementation of the river basin management plan.[18] Agricultural problems detected by field assessments, for example, are referred to specialized agricultural sustainability advisers, who collaborate with the assessment team to help the implicated farmers to improve their land, farmyard, and nutrient management practices.[19] The corps of sustainability advisers links the contextualization of water management at the catchment or territorial level to the contextualization of pollution mitigation measures on the farm, completing the nascent system of local governance.[20]

Three key lessons can be drawn from these experiences with the regulation of agricultural sources of water pollution.

First, penalty defaults and high-level framework legislation—in this case, the WFD—orient initial action and incentivize the creation of new governance instruments for the local contextualization of general policies. Making those institutions work in practice, in particular places, requires the continuing revision of the initial plans in light of experience. The recent flurry of institution building in Irish water regulation—the culmination of systematic investigation and hard experience—was preceded by many false starts and misdirected half measures. Experimentalist governance offers principles of design for these institutions, not blueprints for their construction.

Second, the Irish example illustrates how contextualization supplements rather than substitutes for higher-level decision-making and procedures. LAWPRO review modifies targets set by national and regional reviews, and how and in what order they are approached, but it does not dispense with them. Local authorities and stakeholders are not free to disregard the national framework; local interventions must remain accountable to more general procedures. In this setting, lower and higher levels interact dynamically, correcting and evolving with each other.

Third, these efforts demonstrate the value of blurring the distinction between the regulation and provision of public services. In the Irish case, regulators and farmers frequently work together. Dairy farmers in the catchment projects prepare nutrient management plans with the support of specialist extension agents, who themselves consult with catchment specialists. Farmers with environmental problems respond in collaboration with newly formed catchment assessment teams and a new corps of specialist sustainability advisers. Unlike traditional extension agents, who propagate consolidated expertise, these new specialists—jointly developing improvement plans with individual farmers and each other—reconsider their own understanding as much as apply it. Collaborative investigation is necessary precisely because the current rules and best practices run out; there is no more law to apply because of gaps or other limits in the pertinent texts and jurisprudence. Authorities must improvise instead of looking to codified standards because establishing what should be done goes hand in hand with developing the understanding and capacity needed to do it. The inclusion and empowerment of local actors thus becomes integral to regulatory problem-solving, with potentially far-reaching consequences for the design and operation of government. Indeed, the creation of LAWPRO, as an open-ended addition to local governance, suggests how regulation under uncertainty can reshape democratic participation.

Integrating Renewables on the California Grid

Now we turn to the pivotal industry in deep decarbonization: electric power. Any program of steep emissions cuts will involve massive electrification because it is easier to control or avoid emissions at power plants than at other sources.[21] Though decarbonizing the electric power supply could involve a portfolio of technologies—including nuclear power and advanced fossil fuel plants that capture carbon dioxide pollution before releasing it into the atmosphere—the countries and subnational jurisdictions leading climate policy favor renewables above all other options. Deep decarbonization through electrification has thus become synonymous with the deployment of renewables.

California's experience with the integration of renewables in the grid is a paradigmatic case in experimentalist contextualization. The incorporation of highly variable renewable power sources on a grid changes how the grid itself operates. The size and import of those changes is highly context specific—dependent on the interactions of local conditions with each other and the system as a whole. This makes local adaptation necessary because the economically efficient and reliable use of local resources can only be determined on the basis of actual operating experience. But it also requires the continuing, coordinated adjustment of system-wide rules and supervision to ensure that local modifications do not introduce instabilities into the grid. California has become a world leader in the integration of renewables as well as energy storage and complementary technologies by developing an experimentalist regulatory framework that creates incentives for the rapid, local deployment of innovations, while creating a common language for disciplined exchange and evaluation of experience across levels of the grid system.

As in the case of Irish water regulation under the WFD, however, the road to success was paved with failures. Confident early assumptions about the massive adoption of rooftop solar power were soon disavowed by practical developments; even as it became clear that feasible alternatives would involve much larger units typically under utility control, the regulatory implications of this shift remained elusive because vast new quantities of renewable power required integration on the grid. Conventional designs and grid operations could not perform that task reliably and efficiently; the deployment of novel technologies, such as batteries, would be required. Yet no actor in the system was capable of predicting how these new systems would operate, or which industrial forms and business models would be viable. Incumbents, the utilities, were essential to this process yet also saw threats in the shifting industrial organization. The key California utility regulator, the California Public Utilities Commission (CPUC), felt its way forward—just as CARB did as it transformed the regulation of vehicles in California—ultimately creating new industries and new behaviors by incumbents that led to the pioneering deployment of new technologies while taking account of the demanding requirements of contextualization.

California began promoting renewable energy in the aftermath of the oil crises of the 1970s. State policy, in tandem with the federal government, initially subsidized novel renewable technologies, and created small markets for wind, solar, and geothermal power.[22] Because these new technologies were deployed only at a small scale, it didn't much matter for the grid as a whole how the policies were implemented; the grid simply absorbed the modest, new power supplies. When renewable power ramped up during windy or sunny periods, other generators backed off and vice versa.

The power crises of 2000 and 2001 transformed the politics of renewable power. They showed that traditional strategies for supplying electricity were flawed, opening the political space for new ideas. An amalgam of environmental groups, organized labor, and more conservative political interests concerned about energy security, reliability, and cost sought to diversify the sources of electricity. Renewable power fit the bill. In 2002, legislators set the goal of achieving 20 percent renewable power supply by 2017.[23] As renewable technology advanced at a blistering pace and the renewables lobby gained power, the state set more ambitious goals. By 2018, California was committed to getting at least 60 percent of its power from renewables by 2030, with 100 percent of electricity coming from zero-emission sources (including renewables) by 2045. Such "100 percent clean" mandates, adopted in New York among other states, have become models for planned, nationwide clean energy goals.[24]

These increasingly ambitious goals assumed that power generation technology would keep improving at rates fast enough to keep costs in check. Rapidly expanding global markets produced economies of scale that made this plausible. The *Energiewende*, dating from the 1980s, was subsidizing the rapid expansion of Germany's solar market, making solar energy increasingly attractive economically.[25] The costs plummeted further as production shifted to China from the 2000s.[26] The cost of solar photovoltaic panels per kilowatt of output decreased faster than for any major electric supply technology in modern history.[27]

As the technological frontier expanded, advocates for renewable energy deployment and state regulators largely assumed that meeting California's ambitious goals would be largely a challenge of implementation in the traditional sense: the technologies would advance, and only the local rules and practices would need to be adjusted to allow for full deployment. Delays in the diffusion of the politically most visible systems, small rooftop solar units, did seem to be caused by problems of that sort. In the 2000s, these technologies were thought of as primed to supply half of the state's needed new renewable energy supply, and the state had a goal of covering at least one million roofs with solar—up from just twenty thousand when the law was passed in 2006.[28] But the installation program went slower than expected, and on taking office as governor in 2011, Jerry Brown created a coordinating body to investigate the barriers to the deployment of rooftop technology and codify the best practices in a series of guidebooks—one every two years.[29] The task force first focused on streamlining building codes and permit applications, which were widely known as impediments to implementation.[30] As often happens when government concentrates on implementation in the traditional sense, Brown's task force emphasized the need for coordination of agencies and the clearing of regulatory bottlenecks.

But it soon became clear that putting vast quantities of renewables into service wouldn't just be a matter of regulatory streamlining. Because solar and wind outputs vary with the weather (frequently in somewhat unpredictable ways), they change the properties of the grid, and such changes would in turn require modifications in grid planning, investment, and operation. Moreover, because the grid itself is an interconnected network of regional and local nodes—with electricity on an alternating current flowing in all directions, in ways that depend on frequency and voltage at each node—these shifts would necessitate contextualization on many different levels. By the second installment of the guidebook, in 2014, the process was focused on how local experiences in deploying rooftop solar could guide changes in top-level rules such as interconnection procedures, all with the goal of advancing implementation in the traditional sense.[31]

Ultimately, the diffusion of rooftop technology faced a bigger barrier: cost, and with it scalability. Because such installations were small and decentralized, they remained three to four times more costly per volume of electricity generated than larger, grid-connected solar generators.[32] Being small and decentralized also meant that the electronic gear that connected these generators to the local grid, along with the generator technologies themselves, were less sophisticated than on larger solar systems and less capable of adjusting in real time as rising supplies of solar power altered how electricity flowed on the grid. The more ambitious the state's goals for the overall use of renewable energy, the more it would need to rely on larger solar and wind generators connected directly to the grid. And because California is much sunnier than it is windy, this requirement would entail big solar farms at the scale of fifty to a hundred megawatts (or larger), not a multitude of rooftop systems sized typically at tens of kilowatts.

Bigger systems pumping more power into the grid—serving physically distant customers—would have even larger effects on the grid itself. Rooftop solar installations are designed to be passive and fault tolerant. They trip off if they sense problems on the grid, and that interruption is (for the most part) not itself a further disturbance. To the local low-voltage grid, a house with a rooftop system therefore "looks" mostly like another house—albeit one with an unusual net consumption of electricity on sunny days—and it is invisible to the high-voltage grid for longer-distance power transmission, to which it has only an indirect connection via the low-voltage network.[33] Changes in the output of large grid-connected solar generators, on the other hand, interact in consequential ways with the grid to which they are interconnected; the bigger the system, as a general rule, the more intense the interaction effects that extend fully across the whole high-voltage, long-distance grid. Sophisticated technology is required to ensure that the large power source reacts appropriately to changes in the grid, and especially that these reactions do not trigger additional disruptions. As sources of this type rise in their share of the overall electricity production, the practice of power generation thus shades into grid management.

The grid management problem becomes vastly more complicated than any prior experience managing other grids if the large source of power is renewable.[34] When power generators can be matched in real time and centrally with power usage—as happens when power is provided mainly by fossil fuel or hydroelectric generators, whose output can easily be ramped up or down to meet the expected and real load—it is relatively straightforward to plan a grid system that minimizes congestion on the key power lines and transformers. There are uncertainties, to be sure. For example, it is unknown when a fossil-fueled power plant might trip off-line due to some rare maintenance problem. Grid operators would plan for these uncertainties with a series of simulation models that would mimic the behavior of the grid and contingency plans. If a power plant tripped off unexpectedly, extra units would be sitting idle and ready to ramp up—with the size and location of those backup units dictated by the simulation models.[35] Because these models simulate grids whose operations are highly familiar, they are relatively easy to calibrate with real-world experience. In turn, these models informed and were informed by investments in the grid, such as new power lines, transformers, and sensors that all allow for a more stable flow and the reliable management of the grid. Adding renewables meant output was much less predictable, making simulation and planning for failures harder. Worse, variability in renewable supply is highly correlated among sources: the output of nearly all conventional solar generators peaks in the early afternoon, for instance. The problem of managing large, decentralized, and variable power sources on a single, highly integrated grid are compounded by the fact that most of those solar generators are unavailable when they are needed most to meet the demand for electricity—such as when the sun sets or a whole region is beset with cloudiness.

By the late 2000s, academics had begun to predict the huge challenges that such imbalances in supply and demand would pose for the grid.[36] Early in the day, before the sun rose overhead, there would be a high demand for nonrenewable energy, such as from fossil fuel and nuclear plants. Over the day, that demand would plummet as the solar supply increased and other plants could be ramped down. In turn, demand for nonrenewable energy would rise again in the late afternoon when the sun would start to set, solar generators would taper off, and people still needed lots of electricity. The daily supply curve of nonrenewable generators (the dispatchable generators that grid operators could turn on and off to fill in gaps in the solar output) resembled the bottom half of the profile of a duck in flight, with a small, fat tail in the early morning, a deep midday sag as the belly, and a neck that rose steeply into the evening as lights and air conditioners came on, only to taper off late at night, forming the duck's head and bill, as people went to sleep and the total power demand abated.

California's grid operators knew about this "duck curve" problem, but in the 2010s, as the state began a rapid shift to renewables, they assumed that the biggest challenges in ramping power up and down lay far in the future—in the 2020s and beyond. They assumed that the variability in solar output could be managed readily by adjusting other generators on the grid to fill in supply gaps, and convincing Californians to adjust their demand during the day. They also assumed, as a last resort, that they could curtail any oversupply of power that remained.[37]

To varying degrees, all of these assumptions have proved wrong. The duck curve arrived much sooner than predicted, and managing it has proved harder than assumed in the models and policy plans. These failures to anticipate the speed at which the duck curve appeared also meant that investment decisions for new power lines, storage systems, and other elements of the network hardware that could keep the grid stable lagged behind need. Those investment decisions were, in most cases, the joint product of private investors (e.g., electric utilities or independent power producers) and the two regulators that oversaw the California electric power system: the California Independent System Operator (CAISO), which oversees and operates the high-voltage grid, and CPUC, which regulates the utilities and most of the lower-voltage systems.[38]

The faster the state deployed renewables, the harder it became to predict those realities—yet politically and economically, it remained vitally important for the grid to maintain reliability while integrating renewables at an acceptable cost. By 2017 or so, the imbalance problems were becoming urgent. Curtailment started rising sharply, and that waste implied economic loss; worries about the reliability of the grid rose as well.[39] With the rapid deployment of large, grid-connected renewables, the state was in uncharted—indeed in some ways unchartable—waters in integrating these new power sources into the grid.

Again politics constrained the search for solutions, ruling out the strategy that most engineers favored: enlarging the grid to make it easier to move power as needed, clearing congestion during periods of oversupply, and bringing in new supplies during periods of scarcity. The political problems included siting because new lines were challenging to build. Just one new power line to connect solar generators in the southeastern California desert to San Diego had required nearly a decade to site, approve, and build.[40] Worries about the visible blight from power lines (and fears, unfounded in science, that power lines caused cancer) were the source of enormous delays; tamping them down by moving lines underground would explode the cost tenfold. Moreover, the political coalition that helped cement support for California's shift to renewables included organized labor; California unions liked in-state renewable power because it generated local jobs, and they opposed grid expansions that, for example, would have made it easier to import more wind power from Idaho and Wyoming.

Developments in the battery industry—then and now advancing at a pace even faster than photovoltaics—offered a partial solution to the integration problem. In principle, congested power lines and transformers could be utilized more fully if batteries stored the excess power and reinjected that energy later. This could be particularly important for places in California where congestion was created by excess power that was useless when generated by the midafternoon sun, but extremely valuable just a few hours later. Batteries, along with high-power electronics, could help dampen variations in voltage and frequency, making it easier to manage the grid reliably even as it shifted to renewables. As batteries got cheap enough, the thinking went, it would become less costly overall to upgrade the grid with batteries than to build extra power lines and transformers, and overbuild solar generators whose output would be curtailed during periods of congestion. Politically, this approach would also lower the risk that jobs building solar and wind generators would flow to other states. As many of the world's leaders in advanced batteries, power electronics, and grid controls were based in California—for instance, Tesla—a battery solution could generate economic advantages within the state as well.

While the logic was straightforward, actually using batteries to make the grid more capable of integrating renewables depended entirely on the local context—the moment-by-moment types of congestion experienced on each part of the grid. Translating the rapidly strengthening political coalition in Sacramento and Silicon Valley that favored big deployments of batteries as industrial policy into workable solutions for the grid depended on extensive local planning informed by detailed, current information. Provisional, top-level goals for deploying renewables and batteries could still be set using formal models that simulated the grid. But achieving those goals—and setting new ones that better reflected actual operating experience—required regulators and the deployers of batteries to pay close as well as constant attention to the lessons of practice.

TWO WAVES OF BATTERY DEPLOYMENT

The big push to utilize batteries across the grid began in the 2010s. The first wave of efforts focused on inducing the procurement of the emerging technology by utilities—encouraging them to experiment with the installation of many different kinds of devices in many different settings, thereby driving down the costs and learning the value of various uses. That uptake of devices has been faster than expected, and efforts, still in the early stages, have shifted in a second wave to creating a regulatory framework that encourages a more efficient deployment of batteries yet safeguards the integrity of the network.

The instrument for the first wave had its origins in a 2010 law mandating CPUC to create an energy storage program to support the integration of renewables.[41] The law, only a few pages long, set no targets. Instead it outlined a process by which CPUC would develop statewide goals through local planning. Each utility was required to procure storage systems and document how these storage devices functioned when connected to the grid.[42] This kind of technology forcing under uncertainty recalls CARB's transformation of the California vehicle market in the 1990s, discussed in chapter 4. The big difference is that in the case of CARB, vehicles did not interact with each other or the road network (apart from the need for charging stations), and thus CARB could concentrate more exclusively on the technological frontier and maximizing the deployment of devices on the road. In the CPUC battery case, by contrast, the storage devices and grids very much interact with their context. More devices, alone, would not cause a technological revolution.

Like CARB, CPUC began with an exploratory evaluation of possible uses, focusing on storage systems generally, not just the dominant lithium chemistry. Then (and now) solid-state lithium batteries were thought to be the technological winner, but better rivals might also thrive if given a chance, such as flow batteries that pumped liquid electrolytes through the device. From there, again like CARB, it set a provisional goal (a total procurement of 1,325 megawatts by 2024). That top-level goal included planning targets for each of the state's three utilities, and subtargets for the portion of energy storage systems that each utility would deploy on the medium-voltage transmission system, lower-voltage distribution system, and at customer sites.[43] This approach was designed to ensure maximum efforts at contextualization: each utility would be required to identify novel projects with many different use cases, technologies, and voltages.[44] To underscore the centrality of novelty to the program, CPUC also expressly prohibited the inclusion of any projects that used the one electricity storage technology that was already technologically mature: large pumped hydro.[45] In addition, CPUC required that the utilities contract at least half of their battery projects from independent suppliers. CPUC was creating an industry of the future, and feared that industries of the past (regulated utilities) would not be creative enough to understand and demonstrate creative new battery applications—especially where success might eat away at the traditional regulated utility model.[46] A challenge in creating these industries of the future, of course, was that they would be deploying devices on grids controlled by the incumbents.

CPUC planned procurement rounds every two years, starting in 2014, so that the rules and choices in each round could be updated with local information learned through early procurement. In parallel, CPUC required a comprehensive evaluation of its whole battery storage program every three years, starting in 2016, so that "as more experience is gained . . . lessons can be applied."[47] Just as with the Montreal Protocol's TOCs that evaluated exemptions for critical uses of ODS, discussed in chapter 2, CPUC created a compliance safety valve: a utility could delay meeting up to 80 percent of its planning target by showing that under its current circumstances, the target was for the moment infeasible.[48]

As CPUC turned from planning to implementation, information from the local levels often flowed in much faster than CPUC's carefully planned biennial cycles of updates. For example, in just a year—from 2013, when CPUC set the first quotas, to 2014, when it approved the utilities' first deployments—it became clear from the slate of proposed projects that a new procedure was needed for compensating utilities for the loss of electricity sales from power that was shunted into battery projects.[49] After 2014, as actual procurement advanced, CPUC also learned that the expectations about the likely sites for battery deployment were wrong. At the start of the program, most experts thought that storage capacity would chiefly be deployed on the parts of the grid under direct utility control, including the higher-voltage systems, because that is where electricity flows are greatest and thus the gains from control over these grid-connected system are plausibly the largest too. Instead, many batteries were deployed at customer sites, in part because customers and the equipment vendors operating storage on their behalf quickly learned how to use these battery systems to cut costs as well as improve power reliability. In parallel with CPUC's efforts to advance grid-connected batteries, large power users on their own were buying "behind-the-meter" battery systems that would help them cut electricity costs and make their electric supplies more reliable.

Because the behind-the-meter market was growing so much faster than expected, before it made any decisions about the 2016 procurement, CPUC adopted new accounting methods to allow utilities a 200 percent "ceiling" for overcompliance in deploying projects at customer sites, with overcompliance usable to offset procurement quotas elsewhere on their systems.[50] (As in the sulfur case in chapter 4, what looks on the surface like a market—the trading of compliance quotas—is in fact active, conjoint learning by the regulator and regulated about how to set the rules, and the "trading" that follows is relatively inconsequential optimization within those rules.) At the same time, CPUC adjusted a state subsidy scheme that helped pay the cost of battery installations so that in providing electricity to users that were competing with utilities, known as Community Choice Aggregators, new actors could be engaged more fully in the battery revolution.[51] Once again, the industrial policy of CPUC put it at odds with the utilities, for Community Choice Aggregators were designed to allow communities to take more control over their own procurement of energy services and shrink the revenues that might flow through utilities.

The second wave of contextualization began in 2018. With the conclusion of the second biennial procurement, the state was exceeding its deployment goals for batteries, yet still had not learned much about how batteries would transform grid operations. Under the procurement program, the utilities were required to operate the new equipment in a program-specific experimental zone. Neither the regulator nor the investing utilities knew how procured devices could be put to efficient economic use while also meeting the requirements of the ongoing reliability of the grid. A combination of regulatory mandates and guarantees of fair return on investments allowed for the necessary experimentation while limiting the scale of deployments so that adverse effects on the network would be safely contained. To scale deployment beyond this sheltered zone would require new rules to allow for the efficient use of the technology while continuing to assure network reliability.

The central issue in devising these rules is a shared language of evaluation; as explored in chapter 3, experimentalism requires the ability of those running the experiments and reviewers to understand what is being learned. The utilities that manage their local and regional grids under the supervision of the CPUC, the manager of the statewide high-voltage grid (CAISO), and the private companies that deployed at least half the battery systems all needed the capacity to understand how battery systems might create value and affect the reliability of the grid, and they needed a common language to communicate that understanding.

This language took two forms. One, focused on the benefits from deployment, is "value stacking." Batteries can do a lot more than just store energy. If coupled to advanced software that can communicate with the grid and power users, a battery storage system can perform many other functions or services, from arbitraging time-sensitive price differentials to stabilizing grid voltage and frequency. Some of these services can be supplied simultaneously, while others can only be provided in sequence. In practice, it is as if no single service requires a substantial fraction of a storage system's available resource, and no single service is profitable enough to amortize the cost of investment. The economic use of storage therefore requires rules that allow the combining or stacking of different uses, with the proviso that the combination of committed services not put the stability of the network at risk.

What makes value stacking especially difficult for regulators and storage users alike is that the performance of storage devices is highly context specific, and the exact value from combinations of services was unknowable without experience. As a leading research institute put it in 2015, "The range in value that energy storage and other distributed energy resources can deliver to all stakeholders varies dramatically depending on hundreds of variables. These variables are specific to the location where resources are deployed, making generic approximations of value difficult."[52] It is possible to distinguish adverse uses conceptually (where a commitment to one service jeopardizes serving another) from compatible ones. But the actual performance of particular devices, and hence their value (and risks) to the grid, can be known only through intense and ongoing observation in local contexts.[53]

The other shared language concerns the reliability of the grid and has already been discussed above: grid modeling. Grid operators use a suite of power flow models that simulate how every component (node) of a grid reacts to changes in the behavior of any of the others. Each utility or other local grid operator must maintain its own power flow models that simulate all the nodes under its control, customized to its local setting (with the help of the model software provider) and periodically calibrated to the actual behavior of each grid. These local models are regularly compared to one another under the supervision of the regulators to produce a single, contemporary, system-wide set of best practices for grid modeling even as each grid operator maintains its own grid-specific configuration.

The practice is dynamic and interactive, and begins before any large new devices are connected to a utility's grid, including battery systems. The utility uses simulations of the performance of a proposed device to assess the values of all the services that the device might provide. These simulations of power flow on the grid also identify places where the device might create conflicts or congestion that might require grid upgrades or adjustments to the design of the device itself. Through an iterative process of project proposal, simulation of power flow on the grid, adjustment, and then resimulation, a viable project emerges. With value stacking, the economic value of the project can be assessed. With power flow models, the risks to the reliability of the grid can be assessed. The convergence is formalized in an *interconnect agreement* between the operator of the battery project and the grid operator, stating in effect what the grid in its various layers and nodes can expect from the project under varying conditions, and constraining how the device behaves on the grid.[54] The power flow calculations made by the operator of the grid—in California, that's the utility for medium- and low-voltage grids, and CAISO for the high-voltage grid—are the ultimate arbiter in this process. But the methods and assumptions are transparent so that other third-party contractors can run their own simulators and developers can design projects without always being under the utility's thumb. Along the way, the regulator could check these calculations as it makes decisions such as whether to approve the inclusion of a utility storage device in a utility's rate base or the cost of storage services that a utility purchases from an independent supplier.[55] This is why the concept of value stacking was so important to accelerating investment in battery systems: most of the services provided by batteries have never been valued properly in markets and thus regulators needed a way to determine whether the full range of benefits from deploying batteries could be aligned with the prudent bearing of costs. That interconnect agreement is in turn updated as needed, if only to alert the regulator and grid operator to new power service commitments, or variations in the actual performance of a device, that might eventually interfere with (or complement) other devices or the grid's reliability.[56]

Introducing radically new technologies raises huge challenges for modeling because it is hard to identify the right functional form and assumptions to govern how novel devices are represented. While analogous cases can be identified, the problem is emblematic of the uncertainty discussed in chapter 3—uncertainty that is irreducible in the absence of real-world experience in context.[57] One place to observe this readily is at CAISO, the high-voltage grid operator, because it makes its deliberations around modeling and the procedures for setting interconnect agreements so transparent. As the first battery projects were advancing, each required the modeling of how the battery would affect the surrounding grid and an interconnect agreement. For the high-voltage grid, this meant that the regulator (CAISO) needed to resolve novel questions such as whether batteries were generators or sources of demand, the traditional categories for grid-connected resources.[58] The answer was both and neither, so CAISO created a new category of grid-connected resources, published the code it would use in modeling these resources, and outlined an approach to modeling and interconnect agreements that would be good enough until practical experience might reveal better approaches.[59] As each cluster of projects in California advanced—each needing modeling and interconnect agreements—grid operators also watched as a few other jurisdictions around the world ran experiments in allowing battery interconnection. Until the California market surged in size, making it today the biggest deployer of grid-connected batteries, the largest battery experiment was in Australia, where a large system was installed at the interconnection of two grids. Comparing model-based studies that tried to predict how that Australian battery would operate with what was learned when the battery actually functioned helped reveal how little was predictable reliably in the absence of real-world experiments in system context.[60]

A fundamental task for regulation under uncertainty in this setting is to strike a balance. On the one hand, the regulator is advancing a form of industrial policy; it is pushing for the emergence of a nascent industry and thus wants as much experimentation as possible.[61] On the other hand, it must ensure that the experimentation needed to encourage innovation and deployment does not degrade network reliability or impose unacceptable costs on ratepayers. Thus CPUC's 2018 provisional rules for the deployment of storage authorized utilities to provide stacked services in any combination they think useful, subject to the requirement (in rule 6 of CPUC's "Decision on Multiple Use Application Issues") that "a single storage resource must not enter into two or more reliability service obligation(s) such that the performance of one obligation renders the resource from being unable to perform the other obligation(s)."[62] In effect, rule 6 is a kind of guardrail. Deployers of battery systems are authorized to experiment with value stacking and have strong incentives to find ways to maximize the total value, so long as their experimentation does not involve multiple claims on services that could be needed when the grid is under stress. Neither CPUC nor the battery operator can predict the exact grid configurations or conditions that might affect how batteries perform in real time, but they can control how reliability-related services are offered to the marketplace and contracted. After that, real-world observation is essential to looking, in context, at whether the regulator has struck a balance that is too conservative; if so, the guardrail would need changing.[63]

Periodically, the real-world offers extreme experiences that can test experimental deployments at limits that would be too dangerous to create under normal circumstances and too unusual to simulate reliably. Large-scale grid outages offer a unique opportunity for that kind of testing; at this writing, the most important of these were two days of statewide power shortages in August 2020, a time when hundreds of megawatts of battery projects had been deployed—each operating under interconnection agreements with grid operators that were designed with protective guardrails yet their actual operations would help reveal whether batteries in the context of grid duress could do a lot more for grid reliability than had been allowed. CAISO, an operator of the state high-voltage grid, ran an interagency process after those outages that mapped power flow models to the actual performance of all devices on the grid—looking in particular at whether batteries could be operated in different ways when the context on the grid required more resources.

What CAISO found was that the guardrails were too conservative and led to the underutilization of battery resources in practice.[64] Put differently, the rules had run out. Prior to such real-world experiences it was impossible to set an effective rule that could anticipate all circumstances. Supervision using common languages of assessment, rather than rules, was required—and when the opportunity of extreme conditions arose, that supervision made it possible to write new rules and set the context for a new wave of experimentation in real-world contexts.[65]

In sum, the problem for the regulator and investor in regulated storage is to arrive, initially, at an approximate estimate of the performance of the device that is detailed and reliable enough to warrant a decision on the soundness of the investment, and then correct an understanding of the network with regard to that type of device as local experience accumulates. The common languages of evaluation—value stacking and power flow models that simulate reliability—made this possible. Guardrails such as rule 6 made it safe. This is, again, a paradigmatic case of experimentalist contextualization: local adaptation is necessary (because performance in place can't be adequately estimated without place-based data), but the stability and adaptability of the overarching system depends on continuing the central adjustment to these ground-level changes. Through this process, regulators oversaw a massive deployment of batteries on the California grid—deployments that in just a few years, rose from essentially zero to the largest in the world.

The outcome of this second wave of efforts at contextualization, still underway, is hard to predict.[66] At one extreme, the deployment of storage devices around the grid could unfold in ways that radically reduce the role of the high-voltage grid itself—an extreme vision that is not looking quite so extreme in light of technological and political shifts. Again, the shock of events has altered the politics of experimentation—in this case, the shock created by evidence that power lines (of many voltages) have ignited deadly wildfires around the state. One response has been regulatory incentives for more decentralization. As a follow-on to batteries, California regulators are advancing experiments for clusters of decentralized energy systems.[67] This exploration may or may not produce an additional set of markets for novel, stackable services. But to go by experience so far, it will almost surely clear the way for yet more practical investigation, under regulatory auspices, of more effective power management though more decentralized control, and lesser roles for classic regulated utilities and the high-voltage grid. In 2021, California regulators and policy makers launched, for example, a new wave of experimental investments into microgrids that will link together local battery and renewable energy supplies in ways that could make local electric service more reliable.[68] The challenges for CPUC are identical to those it faced at the onset of the battery program: a nascent industry that in principle, could yield large social value, but whose best configurations and locations for deployment are unknowable without learning through deployment about the context.[69]

Finally, the contextualization of renewables and storage devices on the California power grid has had large political impacts that are reinforcing experimentation. It has fostered, despite the continuing uncertainty, the plasticity of political support for innovation in energy supply and management. Constant change, not stability, creates a sense of possibility, and with it new alliances built around strengthening the political power of novel industries that could not have emerged without the disruption created by new regulatory arrangements. As these novel industries get larger, they fund increasingly powerful lobbying groups that are highly active in state legislative and regulatory affairs—groups that are building alliances with other powerful political actors, such as those that back renewable energy and organized labor.[70] Yet more disruption follows. Now the incumbents, too, are deploying high-profile projects specifically designed to demonstrate how batteries can achieve popular goals such as reducing the dependence on fossil fuels.[71] Since 2010, with authorization from the state legislature, CPUC in effect has orchestrated an industrial policy to nurture an infant energy storage industry, with its financial center in Silicon Valley, and help firms identify and demonstrate growing markets for their products. As of today, that industry is at least an order of magnitude larger in sales as a direct result of CPUC-led procurement, and with that larger size comes more political power. Through disruption, the interests of all the major actors—including all the incumbents—have changed. While disruption was easy to talk about conceptually, what made it a reality was a regulatory system that learned about technology deployment in context. When people speak of a future grid transformed by decentralized electronic devices, including batteries, they tend to focus a lot on how the performance of those systems has improved. But performance has been a function of deployment—and that has moved at the pace that regulators and deployers have learned about context.

Sustainability and Development in the Amazon

For a final example of contextualization, we consider efforts to combat deforestation in the Amazon.

Deforestation in the tropics is a menace to the planet. Among other things, it disrupts the climate, destroys irreplaceable habitats, and displaces communities of Indigenous peoples and other forest dwellers. Tropical deforestation in the Amazon, Indonesia, Malaysia, and equatorial Africa alone contributes about 10 percent annually of the total carbon emissions, making it second only to vehicular pollution as a cause of climate change.[72] Land use protections have a greater effect here than anywhere else. Where Ireland contextualizes the meaning of "good water" in the WFD and California contextualizes the state regulation of novel energy storage devices for the local management of renewable sources on the grid, Brazil is contextualizing the law and regulation of public and private property to give concrete meaning to sustainability as the reconciliation of economic activity and respect for environmental values in the tropics. There is no place where competing ideas of development and its relation to correspondingly diverse concepts of environmental protection are more openly as well as urgently contested among grassroots movements, local and national governments, and national and international NGOs than in the Amazon.[73]

We focus on Brazil because it has embraced the most comprehensive range of measures for protecting tropical forests. Enforcement in Brazil, at times aggressively rigorous, has been the most effective among countries with similar protections, and the resulting successes and failures say the most about what does and does not work. Indeed, Brazil has been called a "laboratory of governance innovation" in these matters.[74] In a first wave of reform, from roughly the end of the military dictatorship in 1985 to the early 1990s, innovations in the use of public land allowed forest-dwelling communities and Indigenous peoples to continue traditional ways that were presumed to have been compatible with the flourishing of the forest over long periods of time. In a second wave, during the two terms of the leftist president Luiz Inácio da Silva, or Lula for short, from 2002 to 2010, the measures directed at traditional forest dwellers were extended, but there was a new focus on commercial land use. Brazil combined the regulation of supply chains in soy and beef with rural land registration, backed up by the application of penalty defaults such as exclusion from subsidized credit or important markets, to induce ranchers and farmers who settled in the Amazon in recent decades to adopt sustainable practices. These measures resulted as intended in a dramatic drop in emissions. But the limits of the reforms with respect to both traditional and commercial land users, exacerbated by political changes, have become clear in recent years and are reflected in increased rates of deforestation. These limits underscore again, and boldly, the zigzag course of these contextualizing reforms, typically over decades, and the need for a background consensus on general goals, no matter how thin, to learn from the failures of partial successes.

The distinctiveness of its reforms notwithstanding, the arc of Brazil's development of the Amazon exemplifies broader tendencies. In treating the tropical forest as unspoiled land to settle and exploit—thus provoking social conflicts and trying to limit their effects—Brazil's experience in the Amazon strongly resembles that of other countries in the forested tropics.

Indonesia is a telling illustration. Brazil shares with Indonesia a colonial past that left a legacy of large estates and a plural legal system in which customary forms of property survive in the shadows of formal, modern law. Both countries have a tradition, anchored in fundamental statutes and honored mostly in the breach, of social property, respecting the place of smallholders in society.[75] Both had a developmental dictatorship in the second half of the twentieth century—the military government in Brazil (1964–85) and Suharto's "New Order" in Indonesia (1966–98)—that brought migrants and corporations into the forest to settle, log, mine, and plant, often in violent collision with each other and the customary communities long there, and all in flagrant disregard of the environment. In both countries, international outrage at deforestation and domestic protest against abuses of power gave rise to movements to make development sustainable. In the end, both movements favored the "greening" of large property, provoking a call for enhanced support for smallholders along with the continuing, place-based contextualization of regulations and property claims.[76] In this larger perspective, Brazil exemplifies the late twentieth-century paroxysm of postcolonial development that has turned much of the tropical forest into a dangerous frontier.

Politics and penalty defaults play a more salient role in our discussion of the Amazon than they have in our other case studies, largely because they more sharply constrain the definition of problems and specification of solutions in the context of deforestation. In the case of commercial land use, penalty defaults are imposed abruptly, almost as they are announced; reforms take shape as the producers, determined to escape exclusion from the market, negotiate what compliance with the requirements of sustainability will concretely mean. We thus concentrate less on a peer review of local experience and other operational routines of experimentalist governance institutions, and more on the preliminary and far less orderly learning that occurs among diverse actors at many levels. Throughout this process, the substance of reform—what sustainability means, practically, in various settings—is contextualized through exposure to contrasting views, first in the rough cast of social movements or the projects of local elites and their allies, and then in the forge of high politics.

PROTECTING FOREST DWELLERS

The military government that came to power in Brazil in 1964 aimed to bolster national security, accelerate the growth of the domestic economy, and improve the lot of land-starved peasants in the south and northeast of the country. Settling the Amazon contributed to achieving all three goals. Settlements, together with military reserves, would obstruct foreign intrusions through the vast, unguarded frontier and limit the use of the territory as a staging ground for guerrilla operations. The slogan was *integrar para não entregar*, or use it or lose it. Given the theories of economic development of the day, the benefits of exploiting the vast store of raw materials in the Amazon were as self-evident as the need for simultaneous investments in national industry to process such inputs. The unsentimental message to investors was *chega de lendas, vamos faturar!* (enough of the legends, let's cash in). Settlement promised to address populist demands for justice in the countryside without the inconvenience of land redistribution. From this perspective, the Amazon was *terra sem homens para homens sem terra*—a land without people for people without land.[77]

The government's initial plan was to attract a hundred thousand families to settle in a fishbone pattern along the planned Trans-Amazonian Highway, transecting the region from east to west, and the BR-163 highway, running north to south. Settlements were planned in regular intervals along the projected roads: a small village every 10 kilometers, an intermediate-sized *rurópolis* every 50 kilometers, and a large city every 250 kilometers. Colonists were to get a homestead, support from agricultural extension services, and access to schools and churches for their families.[78]

But this grand vision proved a bust. The official program drew only thirteen thousand families, and next to none of the promised infrastructure and services were supplied even for them. Only one rural city was built, at the intersection of the Trans-Amazonian and the BR-163. By the mid-1970s, the government changed tack, and encouraged large, corporate investment in logging, mining, and agriculture on a scale to match the national ambitions for industry.

The succession of measures produced a maelstrom of opportunities that sucked capital and people into the Amazon from the 1970s on. Migrants who started ranching went broke, and found work as loggers or miners, hoping to try homesteading again; ranchers occupied the land abandoned by the homesteaders or cleared by the loggers, and converted it to pasture. In an inflation-prone country, investors from the rich south, constantly looking to protect their wealth with holdings in land, financed settlement in the Amazon when the government hesitated. Property titles for the newly claimed territory were generated as needed. Brazil, like other Latin American countries, inherited provisions from Roman law that allowed for the acquisition of property by possession (*posse*) or the homesteading improvement of public land (*usucaptio*). Property claims based on these or other grounds could be certified by the federal and state governments as well as various entities administering settlement programs. Absent a unified, national land registry, certifications conflicted, and land claimed in a given county could exceed its actual territory by half. Taking advantage of this system was a profession that specialized in the production of artfully "aged" property titles that pass for real on first inspection. By the late twentieth century, the end result was to turn the Amazon into a new frontier, with disastrous consequences for the environment.

As all of this was unfolding, the rubber tappers of the northwestern state of Acre, at the border of Peru and Bolivia, were at the forefront of grassroots protests against the new settlers, and these protests grew into a movement that culminated in the creation of a new kind of reserve on public land at the end of the 1980s. Acre had been a center of the rubber boom that drew waves of migrants from the northeast at the turn of the twentieth century. As in many labor-scarce areas dependent on commercial agriculture, the rubber estate owners in Acre immobilized the migrants they recruited by debt peonage, backed up by violent thuggery. For decades rubber tappers resisted this bondage, overtly with periodic strikes and covertly by diversifying their activities in the forests—collecting Brazil nuts, fishing, hunting, and growing food crops for domestic consumption—in order to reduce their dependence on the company store and rubber tapping in general. The industry saw a brief revival during World War II, but as it faded into international irrelevance in the 1960s and 1970s, this diversification became a survival strategy—only to collide with the huge inrush of new landowners, attracted by the state's guarantee of favorable conditions, which restricted the freedom to range in the forest on which the independence of the rubber tappers had come to depend.[79]

Under the charismatic leadership of Chico Mendes in the 1980s, the rubber tappers revived the earlier strike tradition and turned it to the defense of the forest as well as economic survival. The characteristic protest was the *empate*—a pacific standoff in which the rubber tappers literally hugged trees to halt gangs with chain saws. The *empates* went hand in hand with the formation of a national rubber tappers union and a determined hunt for an alliance with the domestic Indigenous movement, itself gathering strength from a worldwide assertion of the rights of First Peoples, and like the rubber tappers, in search of new models of collective or joint landownership.

Early efforts by the federal government to address the problem—by settling rubber tappers on homesteads under the colonization program—only made things worse. The exclusive ownership of a small plot proved as confining to foragers as the pressure of aggressive neighbors, with the additional complication that sales to outsiders by individual members of a tight-knit settlement, in which much land was implicitly shared, could jeopardize the integrity of a whole community.

A solution—the extractive reserve—emerged from discussions among rubber tappers, Indigenous groups, government officials, and domestic and northern NGOs. The idea, the culmination of the first wave of reform, was to lease public land to a community according to a scheme that balanced various interests. Twenty percent of the territory would be assigned to individual families to use as they liked. The remainder would be jointly controlled by the reserve's member families, with access and proceeds divided according either to customary rules or new rules determined by a managing council representing the common interest. Because the reserve area was neither individually held nor collectively worked, it left room for the diversified and changing strategies of the forest dwellers, and because the land remained public, the community was not hostage to the possibility that individuals could sell to outsiders. The goal was to enable reserve inhabitants to continue traditional ways of life, with secure land tenure along with the benefit of such subsidies and technical assistance as might be needed to market their products or develop new ones consistent with traditional practices.

The new type of landholding marked a first, partial contextualization in Brazil of the ideas of sustainability and development. Against the prevailing "fortress" orthodoxy of northern NGOs, by which protection of vulnerable ecosystems required the nearly complete exclusion of humans, the creation of the reserves allowed that humans could coexist with tropical forests so long as their activity makes them natural stewards of sustainability.[80] Arriving at this "socioenvironmental" or "neotraditional" understanding of permissible land use in the Amazon cemented the relation between "big" (global in its reach, and formal and explicit in its methods) and "small" conservation (local in scope, informed by individuals' day-to-day choices, and often based on traditional or tacit knowledge invisible to outsiders).[81]

IMPOSING PENALTY DEFAULTS ON RANCHERS AND FARMERS

But even as forest-dwelling communities gradually came under the protection of this and similar arrangements, the pressure on the forest from ranching and farming increased, and with it the pressure for reforms specifically aimed at regulating these commercial activities. Extreme macroeconomic instability—including high inflation in the first half of the 1990s, a sharp revaluation of the currency to reduce it, and a recession ending in a full-blown debt crisis at the end of the decade—did not deter investment in larger and larger infrastructure projects. Hard times and the absence of alternatives made the Amazon a haven, and waves of settlers crowded in. At the same time, the Amazon, already connected through the export of spices and rubber to long-distance trade, was becoming ever more visibly part of the global economy, with predictable effects on the environment. The rate of deforestation, calculated annually on the basis of satellite imagery by the Brazilian Space Agency since the end of the 1980s, spiked in the mid-1990s, and then began to climb again, tracking, with a lag, the movements of the prices of beef and soy.[82]

The government responded to such alarms mostly through changes in the law that, unenforced, only signaled concern without requiring action. The fraction of land that rural property owners in the Amazon had to leave undeveloped was increased from 50 to 80 percent. Owners likewise were required to maintain protective buffers on riverbanks and steep slopes. Criminal sanctions could be imposed for violations of the environmental requirements; corrupt officials complicit in such violations were made subject to criminal sanctions as well. A new kind of conservation unit was created—in one variant, prohibiting development entirely such as in natural parks, and in another, allowing only sustainable uses. The most tangible response was the creation of additional extractive reserves on the lines of the model developed in Acre.

But in part because of muffled official responses to deforestation, northern NGOs redoubled their engagement in Brazil in these years. The Rio Earth Summit of 1992, organized by the United Nations, drew world attention to the threat to the Amazon and elsewhere, but familiar disagreements—over responsibility for the danger and financing of solutions—combined with the constraints of consensus decision-making disappointed northern hopes for a binding international regime to limit environmentally unsustainable logging. The World Wide Fund for Nature began to fill the institutional void in 1994 by bringing together forest owners, timber companies, forest dwellers, and environmental NGOs in the Forest Stewardship Council to oversee the development of a code of good practices for the entire global forest products supply chain. Certified adhesion to the code would distinguish good actors from bad ones and expose the latter to penalty defaults enforced by rich-country customers in international markets. Later, many similar efforts would follow to use supply chains to extend the reach of rich-country norms to sectors implicated in deforestation, such as palm oil and soy.

In Brazil, Greenpeace pursued the same goal more directly. Without first establishing a code of acceptable conduct, the organization simply identified illegally harvested timber and denounced the reputation-sensitive companies that exported products made from them. In the 1990s, the Greenpeace campaigner sneaking through the damaged forest to mark contraband logs with luminescent paint for tracking thus replaced the tree-hugging rubber tapper as the symbol of environmental activism.[83] These efforts were the seeds of national supply chain agreements in soy and beef that over the following decade demonstrated the strengths and weaknesses of such supply chain regulation before they were exposed elsewhere.[84]

The presidential election of 2002 brought environmental issues to the fore in national politics, as Lula and his Workers' Party finally triumphed after two other bids for office. Lula had been allied with Mendes and the rubber tappers since the 1980s. The one secretary in the Ministry of the Environment retained from the outgoing government was Mary Allegretti, who had worked closely with Mendes and the NGOs in developing the idea of extractive reserves. The new minister of the environment was Marina Silva, a close confidant of Mendes's in the labor movement in Acre and then representing the state in the federal senate. The governor of Acre had likewise been a confidant of Mendes. The expectation was that the new government would shift from defensive measures assuaging foreign and domestic concerns about deforestation to an active defense of the environment and forest dwellers.[85]

But support for environmentalism was qualified. Like other "pink tide" leaders coming to power in Bolivia, Ecuador, and elsewhere at the time, Lula combined a concern for Indigenous peoples and their home ranges with the conviction that natural resources are the patrimony of the nation, to be exploited, to the extent economically feasible, for the benefit of all.[86] To this view was added the qualification, especially pronounced in Lula's case, that the proceeds of natural resource exploitation were to be primarily used for redistribution—to reduce poverty and compensate vulnerable groups for harms suffered, including harms connected to the extraction of resources—and only secondarily to alter the structure of production itself by investing in new industries or less harmful ways of organizing established ones. The costs to the defense of the environment along with the growth prospects of the economy as a whole would only become clear as the commodity boom that financed redistribution ran out and the political winds shifted abruptly.

A 2003 Brazilian Space Agency report of a spike in deforestation, back to the mid-1990s' peak, triggered the government to create the Action Plan to Prevent and Control Deforestation in the Amazon. Initially the plan was for the period 2004–9. Through successive extensions, it remains, though much battered by recent governments, the framework for policy today. In an exceptional move for Brazil, the program reflected a whole-of-government approach: the plan was to be overseen by a council of thirteen ministers from key areas, and administered within the Casa Civil or Executive Office of the President (until 2013). This made possible the rapid and effective coordination of measures ranging from criminal prosecutions of corrupt environmental officials to credit restrictions imposed on producers in supply chains or located in municipalities implicated in deforestation. The focus at the beginning was on the "arc of fire" menacing the Amazon along its southeastern border and the disruptions caused by BR-163.

One of the principal goals was the extension of the various types of environmentally protected areas in the Amazon. At the urging of a grassroots network of small farmers' and settlers' organizations on the Trans-Amazonian Highway, including national and international NGOs—the same constellation of actors as in Acre, and like them, fusing big and small conservation—the government approved the creation of a 5.6-million-hectare mosaic of reserves, including Indigenous lands, in the Terra do Meio (land in the middle), between the Xingu and Iriri Rivers in Pará.[87] By the end of Lula's second government, in 2010, some 45 percent of the Amazon was under legal protection, half in Indigenous territories and half in conservation units created by the National System for Protected Areas in 2000.[88] The internal governance of many of these areas remains in disarray, leaving them vulnerable to changing winds, but by international standards this has proved a remarkable effort to protect the environment.

The second major goal was immediate: the maximal enforcement of environmental requirements. In experimentalist governance, severe penalties such as market exclusion are typically imposed as a last resort—when an actor repeatedly proves unable or unwilling to meet a standard, despite support. Under the Action Plan to Prevent and Control Deforestation in the Amazon, the sequence was reversed: the penalty default of exclusion was imposed almost from the start, touching off a frantic search for feasible ways of meeting the regulatory requirements by the producers, the government at various levels, and NGOs, whose own, parallel efforts to control deforestation through campaigning led them to develop promising forms of compliance. The resulting innovations were then fitted together into the second set of contextualizing reforms, focused on the regulation of commercial activity in the Amazon.

The technological key to rigorous enforcement was the improved satellite surveillance of deforestation to permit near-real-time monitoring. The authorities were now able to intervene while illegal activity was in progress, and equipment and contraband could readily be connected to crime. The seizure of goods was significantly more effective as a penalty and deterrent than the imposition of fines, which were notoriously hard to collect. The improvement of the technology was accompanied by an improvement in the direction and coordination of interventions. Central to the enforcement efforts was the Ministério Público Federal, an elite core of prosecutors created by the constitution of 1988 as independent from all branches of government, and charged with defending citizens' rights too "diffuse" or "collective" to be vindicated by the normal legal process.[89] Acting together with their offices in the states, Ministério Público Federal prosecutors were especially effective in bringing charges against corrupt officials in the environmental administration, who were often shielded by local political interests.

The combination of real-time information and enhanced coordination produced explosive—and to all appearances, unplanned—results. The Federal Police swooped in to catch deforesters red-handed, confiscating vehicles, farm equipment, and thousands of cubic meters of illegal logs or herds of cattle. Ranchers and farmers burned police vehicles and administrative outposts in turn, while the federal authorities prolonged raids into "field" interventions, temporarily occupying hot spots. In 2008, the government focused its efforts on a list of thirty-six (soon to be fifty) "priority" target municipalities, chosen on the basis of cumulative deforestation and high rates of burning in the preceding three years, and together accounting for more than 40 percent of the forest loss in the Amazon at that time. All rural landholders within the boundaries of the target municipalities stood to lose access to subsidized agricultural credit and large customers as well.

It was in managing these sanctions, and especially in determining the conditions that priority municipalities would have to meet to be removed from the list, that the government's efforts at enforcement fused with the approach to compliance that had emerged in negotiations among NGOs, large farmers, agribusiness, and state governments in campaigns for environmentally responsible agriculture in the same years.

INVENTING COMPLIANCE

The cascade of innovative improvisations began in 2006, when Greenpeace blockaded the harbor of Cargill's soy-processing facility in Santarém. In anticipation of this, The Nature Conservancy had entered into discussions with Cargill about the possibility of building a system for monitoring suppliers' environmental compliance, property by property.[90] This proposal became the template for a soy moratorium, which endures to the present, under which suppliers that ceased deforesting by 2006 were permitted to continue to sell grain to large domestic and international traders, provided they put their property under regular environmental scrutiny and did not clear more land.

The soy moratorium drew the attention of Lucas do Rio Verde, a rich municipality in the Cerrado (the savanna of Mato Grosso), politically dominated by representatives of large agribusiness firms and soy farmers. To avoid guilt by association with deforestation efforts, the mayor proposed the creation of a land use map in the entire municipality and offered assistance to individual property holders to register under it. As bad behavior by one landholder could damage the reputation of all, environmental law became implicitly a collective responsibility. Landholders were also encouraged to take advantage of a state law allowing them to acknowledge deficits in land use, such as a failure to fully maintain the preserved areas required under the Forest Law, and pledge remedies within a grace period. Making such acknowledgments and pledges established eligibility for regulatory approval to engage in environmentally sensitive activities, including legal land clearing. When the federal agency responsible for environmental enforcement sued landholders in Lucas who, relying on the promised forbearance, had reported infractions, Mato Grosso created a comprehensive, georeferenced rural environmental cadastre for the state—the Cadastro Ambiental Rural (CAR)—and made registration a requirement for environmental permitting.

This agreement—the permission to continue trading in the market in return for a carefully monitored pledge to stop deforestation and eventually correct the environmental deficits—became key to compliance for the priority list of municipalities. To be removed from the list, a municipality had to reduce deforestation to less than forty square kilometers a year and demonstrate that 80 percent of the rural land within its boundaries was registered under CAR, which was quickly adopted in Pará and nationally in reform of the Forest Law in 2012. The aim was to forgo the punishment of wrongdoing in the past, and instead insist on continuing scrutiny that allows for the quick detection of wrongdoing and also increases the state's ability to plan as well as oversee environmental measures. The result features the rudiments of an experimentalist monitoring regime. An additional benefit was to decouple decisions about regulatory permissions and creditworthiness from the vexed clarification of property titles. (The Terra Legal program of 2009 addressed this latter problem, primarily by helping smallholders acquire secure claims through adverse possession of public land. But implementation proved difficult because of the many administrative complications that had thwarted earlier property reforms.)[91]

The same principles were rapidly applied to regulate the supply chain for beef in two, overlapping agreements in 2009. The first was between the Ministério Público Federal office in Pará and meat-packers and slaughterhouses; the second was between the most important slaughterhouses and Greenpeace. The Ministério Público Federal in Pará sued ranchers for illegally deforesting and their slaughterhouse customers for purchasing contraband cattle; it also threatened to sue supermarkets, the slaughterhouses' large retail customers. Besieged, the meat-packers in Pará entered into Terms of Adjustment of Conduct (*Termos de ajustamento de conduta*) with the Ministério Público Federal, suspending prosecution so long as the slaughterhouses did not purchase from noncompliant producers, and the ranchers registered their holdings under CAR. These agreements put meat-packers, in particular, under the continuing oversight of the Ministério Público Federal, and they soon spread to two-thirds of the federally inspected slaughterhouses in the Amazon. A few months after the Terms of Adjustment of Conduct were formalized, Greenpeace created an even more demanding "zero-deforestation" regime under which Brazil's then four-largest meat-packers—Marfrig, Minerva, JBS, and Bertin—agreed to purchase exclusively from CAR-registered suppliers that stopped *all*—not just illegal—deforesting, with compliance monitored by a purpose-built tracking system based on Brazilian Space Agency data.[92]

The combined effect of government and NGO pressure on soy and beef supply chains, together with the imposition of penalty defaults on priority municipalities, was a drop of 70 percent in annual deforestation in the Amazon between 2004 and 2014. Even careful observers concluded that such a large decline showed that "it is possible to manage the advance of a vast agricultural frontier."[93] But given the shock to the world economy from the financial crisis starting in 2007, questions emerged about how much of the fall could be attributed to policy measures. Studies taking into account changes in prices, exchange rates, the size of herds, and the area under cultivation, among other factors, did find that the policies, in total, had a significant effect—a result also corroborated by studies of particular policies.

A comparison of listed and unlisted municipalities, for example, found that listing produced a significant decline in deforestation through monitoring and increased enforcement.[94] A detailed investigation of the Terms of Adjustment of Conduct agreements in Pará showed that, after the agreements, buyers stopped dealing with ranchers engaged in deforestation, and ranchers with whom they continued to do business registered with CAR well before their neighbors outside the new supply chain.[95] The restriction of subsidized credit, imposed only on farmers and ranchers unable to demonstrate the intention to comply with environmental regulations, led to a sizable drop in credits to large ranchers and was associated with a significant reduction in deforestation.[96]

In light of these results, there was new confidence that the state and civil society could work together to design and implement effective policies for reducing deforestation. In 2008, Norway sponsored the creation of the Amazon Fund to finance projects meeting agreed-on reduction targets, and interest was revived in a UN program, Reducing Emissions from Deforestation and Forest Degradation (REDD+), which pays developing countries for results in reducing or eliminating carbon emissions.[97]

THE LIMITS OF REFORM

But this confidence did not last. As it turned out, these years marked the apogee of hope in these kinds of no-nonsense measures, not the beginning of a new age of self-assured management of tropical forests. Deforestation rates in the Amazon have ticked upward since 2012, though they still remain well below the peaks of the early 2000s.[98] Political changes have played an important role in this pattern. The current Brazilian president, Jair Bolsonaro, an apologist for the military dictatorship and its most unsparing ideas of developing the tropical forest, has all but declared open season on environmentalism and Indigenous peoples. Stark as they are, however, these attacks have only exacerbated and laid bare important flaws present all along in the regulation of commercial land use as well as the strategy of supporting forest dwellers by directing them to neotraditional activities.

For example, the long-term viability of traditional activities in the extractive reserves depends on both economic and social conditions: inhabitants must be able to make an acceptable living and also find such activities fulfilling. The more meager the returns to neotraditional activity relative to accessible alternatives, the greater the risk that it is discredited and regarded as inferior in comparison to more lucrative livings, making its economic shortcomings even less tolerable.

Such devaluation is precisely what unfolded over time. By the mid-1990s, only a few years after the new property form was legally formalized, there were already signs of such a spiral of disaffection in the reserves. A study of two distant reserves, for example—Uruará, a municipality on the Trans-Amazonian Highway in eastern Pará, and Xapuri, in Acre—found increasing cattle ranching by smallholders in both regions. Rubber tappers had traditionally kept some dairy cows as part of their subsistence diversification strategy, providing perhaps 4 percent of their annual income. But the rapid growth in the market for beef, better control of foot-and-mouth disease, and improvements in productivity through better pasture management made commercial ranching attractive to them for the first time. Beef prices were higher than those for annual crops and less volatile than the prices of perennials, and once food-and-mouth disease was brought under control, ranching was less exposed to pests than farming. Cattle also proved attractive because of their liquidity—they are marketable year-round—and as a store of wealth.[99] By 2010, an anthropologist doing fieldwork in Acre could see a cowboy culture, replete with the insignias and rituals of flashy belt buckles, rodeos, and barbecues, literally erasing signs of the rubber tappers' struggles as billboards displaying Mendes in the state capital were painted over with ads for hamburger stands and cell phone carriers from one week to the next.[100] The conflicts have recently broken out into the open, as deforestation rates shot up in the Mendes reserve and a dissident group campaigns to secede.[101]

But if this was the outward sign of the cultural and economic failures of the neotraditional policy, the inward and ultimately more consequential marker of disarray was the breakdown in internal governance in the supposedly self-governing areas. Only half of the reserves and conservation units are estimated to have management plans directing development, and only 45 percent have a representative management council charged with making such plans.[102] As collective projects and the solidarity underpinning them came under strain, the will and instruments to adapt as well as renew earlier commitments were failing too.

The limits of using penalty defaults to incentivize registration under CAR—and with it, a shift to sustainable practices—were also evident almost from the start. The agreements in Santarém and Lucas do Rio Verde establishing CAR involved local economies specialized in soy as well as the elites of large rural landowners and representatives of agribusiness already committed to shifting from extensive, low-productivity methods premised on cheap land to higher-productivity, intensive methods based on technology and skill. Compliance with the new requirements accelerated developments already underway, without an abrupt change of course. The same was true of Paragominas, a prosperous, consolidated farming and ranching municipality in the northeast of Pará, which became the first priority municipality to exit the list. Its success was credited to a local pact among the municipality and producers' and civil society groups that created a forum and monitoring structure for the support of registering rural property.[103] The state government, explicitly acknowledging that increased enforcement alone would not secure continuing compliance, embraced the idea of a "green municipality" and made the formation of such self-governance institutions a condition for leaving the priority list. For large, well-capitalized ranchers and farmers in these circumstances, compliance meant a kind of reprieve in which a prior record of deforestation or fraud no longer cast a shadow on current dealings, and public officials recognizing good faith commitments to sustainability became allies rather than opponents.[104]

Where the shift to intensive methods was incipient, though, as was typically the case with smallholders, and the political and social conflicts among groups in the municipality were pronounced, penalty defaults without complementary support measures could not compel compliance. São Félix do Xingu, a municipality at the eastern edge of the Terra do Meio, became a paradigmatic case. In the 1990s, its cattle herd grew vertiginously to become the largest in the country. This expansion drew large ranchers from the center-west and south, and some ten thousand poor smallholders (just under 90 percent of the holdings, accounting for 20 percent of the land), mostly from the northeast. The smallholders—often veterans of failed settlement programs or stints in mining—hoped to earn enough to expand, but were ready to sell their cleared land to larger owners in hard times. Property disputes were rampant. Of the hundred-largest suspect land claims in Pará during this period, thirty were in São Félix. Political power changed hands frequently; there was no recognized elite to direct responses to the municipal listing.[105]

The Nature Conservancy, which continued its series of exploratory interventions, recognized quickly that achieving compliant sustainability in São Félix required interventions to support smallholders.[106] Together with the municipal government, it applied to the federal Low-Carbon Agriculture Program for training and subsidized finance for sustainable methods; a second initiative helped smallholders diversify production and reforest degraded areas by shifting into cacao agroforestry. There were also hopes of finance from the Reducing Emissions from Deforestation and Forest Degradation system linked to the Paris Agreement. But these and other measures failed, and São Félix stayed on the priority list.

Normal accidents of implementation were partly to blame. The federal government rejected the municipality's reasonable contention that the forty-square-kilometer ceiling on annual deforestation should be adjusted to reflect São Félix's extraordinary size (four times that of Paragominas). The NGOs focused disproportionately on the virtues and mechanics of participation, at the expense of discussions of plans and projects, heightening expectations only to quickly disappoint them. But the deeper cause was simply the gaping discrepancy between the smallholders' needs and the available resources. State extension agents had no funds to go to the field, and there were far too few of them in any case. There were reportedly only three municipal tractors for use by all ten thousand smallholders in São Félix; three trucks mustered by the federal colonization agency were broken down in the garage. The state agricultural secretariat had no projects in São Félix.[107] In this de facto disregard for smallholders, São Félix was not an exception. In Paragominas, which was richer, they were neglected in exactly the same ways.[108]

Left to their own devices, the smallholders struggled to adjust. Distressed owners sold out to better-placed neighbors. Land concentration increased. Charcoal production for domestic and commercial use and other rural trades was effectively prohibited. Real GDP per capita declined in São Félix following the 2008 listing, while it increased in Paragominas, Lucas, and Santarém, where less adjustment was needed and, given their wealth, the burden was easier to bear. The rates of deforestation in São Félix ticked upward to double those in Paragominas (but only a fraction of earlier peaks) as some chose flight forward, deeper into the forest, over adjustment in place or a retreat to wage labor.[109] For this reason, smallholders' role in deforestation in the Amazon generally increased in the following year.[110]

For a time, the same welfare measures by which Lula and his Workers' Party (PT) won the allegiance of the poor in Brazil provided a buffer to smallholders, residents in the extractive reserves, and other groups in the Amazon against the harshest consequences of the new policies.[111] Between 1996 and 2006, the share of various social welfare payments in the total family income on the Mendes reserve, for example, increased from 25 to 44 percent, while the share of earnings from neotraditional activities declined from 35 to 9 percent.[112]

Yet as commodity prices slumped, and budgets for welfare and other subsidies were cut, the buffer effects subsided and political support declined apace. In a pattern now familiar in the advanced democracies (and returning to a populist tradition well anchored in Argentina, Peru, and Brazil itself), Bolsonaro's scornful rejection of the status quo revealed a discontent that in retrospect, seems to have been always and unmistakably there. In the Amazon, his election in 2018 amounted to a repudiation of recent policies. Leaders of the producer associations and a trade union operating inside the Mendes extractive reserve estimate that 70 percent of the inhabitants voted for Bolsonaro, who campaigned with the promise to "let the whole of the Chico Mendes Reserve be deforested to build ranches." The leftist coalition that had governed Acre since 1999 and actively supported the sustainable, neotraditional forest economy was voted out.[113] The state capital, where Silva got her political start as a council member, voted for Bolsonaro. Similar results were repeated across the Amazon.

REBUILDING ON THE SAME FOUNDATION

What are we to make of this mixed legacy? As we see it, the political collapse of deforestation efforts, difficult as it will be to reverse, overstates the failure of environmental reforms in the Amazon just as much as the drop in emissions following the priority listings overstates its successes. The forms of collaboration that allowed for rapid learning a decade ago remain models for further innovation, even if the social movements with which they were originally associated are now greatly weakened.

In an effort to repair the social fabric within the reserve areas, new projects are recruiting new generations. The aim is to reconstitute the organs of self-government—a precondition for further development.[114] The grand principle of the reform of commercial land use—amnesty for past behavior in return for a monitored pledge to cease illegal activity to eventually correct environmental deficits—has proven its worth in the regulation of medium and large property.[115] It failed where it was most needed and scarcely tried at all: in supporting smallholders to adjust to the demands of sustainability. Rededication to this principle, aimed squarely at the inclusion of smallholders, could consolidate the gains of the earlier reform. By expanding the gamut of economically and environmentally sustainable possibilities for small property, it could also suggest novel ways—as unanticipated today as the idea of the extractive reserve was three decades ago—to free forest dwellers from the constraints of neotraditionalism while continuing to use the public ownership of land as a safeguard for the protection of an invaluable and highly vulnerable biome.

Many of the background conditions for such a rededication are already in place, starting with smallholders' disposition to give offers of adjustment support a chance. Smallholders at the forest margin know the fragility of the soil all too well and see the economic rationale for sustainability. In São Félix and elsewhere, the rule of thumb is that migrant settlers stay in place for ten or fifteen years before declining productivity forces them to move on. A decade ago in São Félix, moreover, many producers, large and small, were already beginning to adopt sustainable farming practices, improving pasture management, recuperating degraded areas, and stopping deforesting. About 75 percent of those interviewed in a small study said they made changes for fear of being fined otherwise, but about the same percent of the interviewees said they acted to improve returns in degraded areas where productivity was falling, or more directly to "learn from our mistakes."[116]

This experience explains why the smallholders in São Félix were at first enthusiastic about the prospect of working with NGOs and extension agents on more ambitious projects, only souring on cooperation when nothing came of it. Practical experience of this kind informs daily decision-making, often in disregard of contrary political convictions or calculations of immediate advantage.[117] For example, in Rondônia—a state with a weak enforcement regime where deforestation has long been tolerated—a recent study finds that despite the improbability of sanctions, smallholders prepare plans for land restoration, especially when supported by extension agents.[118] The chances are good, in other words, that a substantial group of smallholders would take up a credible offer of adjustment support, and their initial participation could begin the progression from thin to thick consensus at the municipal level thwarted the first time around.

Nor do there seem to be significant technical barriers to raising the productivity of smallholder ranching while making it more sustainable. It should be possible, using available methods, to raise the stocking rate from one head of cattle per hectare to three, and decrease the time needed to fatten cattle for market by using better pasture grasses and management.[119] Smallholder ranching can also be combined with cacao growing, as planting cacao helps restore degraded soils, and the income from the crop shelters the rancher against fluctuations in the price of beef. Pilot programs demonstrating the economic viability and environmental sustainability of this strategy are well established.[120] There are underexploited possibilities for innovation in the use of traditional products such as latex.[121] Brazil's share of the world markets for agroforestry, aquaculture, and fishery products exported from the Amazon is surprisingly small, even when compared to competitors at the same level of economic development.[122] For example, Brazil, which once dominated the export market for the eponymous nuts, now lags behind Bolivia, which has kept abreast of changing international food safety standards while producers in the Brazilian Amazon have not.[123] There is, in sum, room to grow by adopting demanding but proven practices.

Nevertheless, should there be technical obstacles to less land-intensive, more sustainable practices, Brazil's agricultural research agency, Embrapa, is well equipped to address them. Founded in the 1970s as a publicly owned corporation, Embrapa is widely admired for innovations that make the Cerrado arable, greatly increasing the yield of native pasture grasses by crossing them with African varieties, and innovations in sugarcane growing and processing crucial to the success of Brazil's biofuels program.[124] Adjusting the existing solutions to the needs of smallholders, or finding new ones—including, for instance, crop complements to ranching beyond cacao—seems well within its reach.

The chief obstacle to smallholder adjustment is the absence of extension services, as the experience of São Félix again makes clear. Brazil created a federal extension service when Embrapa was formed, but it fell victim to the austerity measures of the 1990s and the general skepticism of donor organizations such as the World Bank at the time that the state could effectively provide support services to private firms. Efforts to reconstitute extension under Lula led nowhere, and today such services, where they exist, are typically provided by chronically underfunded state agencies. From 2000 to 2016, the divorce between policies for family farms, on the one hand, and those for large landowners and agribusiness, on the other, was openly embodied in the coexistence of the Ministry of Agriculture, Livestock, and Food Supply, serving the latter, and the Ministry of Agrarian Development, serving the former. The dissolution of the Ministry of Agrarian Development in 2016 and its reconstitution as a unit of the Ministry of Social Development underscored what had been largely the case from the beginning: support for small, family farming is often considered more a matter of social welfare than productive development.[125]

Correcting this deficit and building an extension service—more precisely, one that can jointly develop adjustment strategies with smallholders themselves—is largely a matter of political will closely tied to the national prospects for development. Brazil has accumulated significant experience in recent decades with programs providing support services to firms—successfully in sectors such as aeronautics, unsuccessfully in sectors such as capital goods for the nascent offshore oil industry, and with mixed results in helping to revitalize clusters of small- and medium-sized firms producing shoes, furniture, and garments. The smallholder sector in agriculture is vast and highly differentiated; 36 percent of Brazil's 5.17 million farms have less than five hectares, and small- and medium-sized firms up to a hundred hectares produce more than half the food consumed domestically, supplying a key assurance of food security.[126] Its significance to the overall growth prospects is increased by the fact that Brazil is among the many middle-income countries in Latin America, Africa, and Asia in which industry's share of the national income has stagnated or gone into decline. At the moment, there are few good alternative employment opportunities for the low-productivity rural population. To accept stagnation in this sector is thus to mortgage growth. Under these circumstances, the failure to find the political will to develop extension services for smallholders is not so much a failure to reconcile development with sustainability as one of development itself.

CONVERGENCE WITH INDONESIA: THE JURISDICTIONAL APPROACH

As we noted at the outset, Brazil is not alone in facing this challenge. Indonesia, to take only the most prominent example, has come by a distinct but similar path to the same crossroads.

In Indonesia, the chief threat to the tropical forest is from the cultivation of palm oil, the cheapest and fastest-growing source of vegetable oil for wide use in cooking as well as an ingredient in processed foods, cosmetics, and biofuels. As in Brazil, smallholders are family owners mixing commercial and subsistence activity on plots that are small relative to those of the large neighbors that hem them in. In Indonesia, too, smallholders frequently encroach on state forests, and as of 2014, they account for about 40 percent of the national palm oil production. Land conflicts are common and increasing, especially those between large commercial estates or plantations and the most politically marginal groups with the weakest land claims. As in Brazil, smallholder yields are low compared to best-practice benchmarks (40 percent lower); yields of commercial operations can be twice as high. The difference is again due to the smallholders' reliance on inferior planting materials and management techniques. In Indonesia also, recent regulatory changes, including a mandatory requirement that all farms and firms in the palm oil supply chain comply with a national code of environmental conduct, impose burdens on small producers that they will unlikely master without substantial assistance—even as those regulations create, as in Brazil, a legal safe harbor for large landowners.[127]

Unsurprisingly, then, proposals for continued reform in Indonesia converge with those in Brazil in seeing enhanced, place-based governance as the crucial instance for linking current land use and supply chain regulation to the construction of a support system for smallholders. The starting point of this "jurisdictional" or "territorial" approach is the recognition that local government is too often ignored in supply chain measures and land reform. Given the heterogeneity of the circumstances, place-based regulation should be at the center of a network coordinating the efforts of local regulators and service providers, farmer organizations, and producer associations and NGOs. As in the LAWPRO reforms in Ireland, the aim is to make environmental sustainability routine and routinely effective by institutionalizing joint problem-solving among the implicated ground-level actors, public and private. Because the green pacts formed in response to priority listing took the municipality as the operating unit, and demonstrated both the potential and pitfalls of inclusive participation, they are seen as a promising model for further development.[128] An extensive network of NGOs is conducting pilot projects under the rubric of the jurisdictional approach in Indonesia, Brazil, and a number of other tropical forest countries.

This convergence can be taken as a first and limited corroboration of the plausibility of the approach. Barring a merciless campaign of expelling smallholders from the countryside at incalculable environmental, economic, and social costs, the alternatives for the tropical forest generally are the same as those for Brazil: inclusive, sustainable growth, or a grinding stalemate at the expense of both growth and the environment.[129] The lessons of recent experiences—both the surprising successes and the limits—is that urgent efforts to avoid the worst could well produce an outcome better than we may have dared to hope.

Innovation in Place

Taken together, these three case studies of contextualization are reason for optimism. They provide evidence that the institutional conditions needed for both technological innovation and political accountability are falling into place in key sectors in countries leading global warming policy. They also show experimentalist institutions coming to grips with three broad areas of the economy—agriculture, electricity, and forestry—that must make the most profound changes in technology and behavior if deep cuts in emissions are to be achieved.

While the details differ across the cases, three common themes emerge.

First, no matter the sector, political pressures and technological changes lead to profound technical, economic, and social uncertainty that must be addressed locally. In agriculture and forestry, uncertainty arises from new political pressures for a radical reduction in environmental footprints. In electric power generation, uncertainty arises from a combination of political pressures along with an unexpectedly rapid change in the technologies for generating and managing renewable power that create completely novel configurations of power flow on electric grids. General responses to this uncertainty—regulations fixing nitrogen limits or acceptable forms of battery interconnection; forest codes; or set forms of using public land—simply do not work out of the box, or at all, when applied to the particularities of local circumstances. Contextualization is needed.

Second, uncertainty creates strong pressures to experiment with new technologies, business practices, and regulatory frameworks, especially at the local level. Pollution control, and deep decarbonization more generally, requires close and continuing attention to what happens on the ground. As we have stressed, what works in one place must often be adapted or reinvented to work in another. Given the dependence of modern production systems—whether agricultural or industrial—on the rapid detection and correction of errors, efforts to improve economic performance likewise demand unswerving attention to context. In order to make clean production economically feasible, deep decarbonization will have to follow these models in laying new emphasis—by public and private actors alike—on local decision-making. In Brazil, municipalities and states learned from each other as innovative ideas of how to comply with environmental penalty defaults proved their worth; in Ireland, local experience is pooled in catchment area programs designed with that purpose; and in California, the state regulators socialize the results of local projects. Governance, no less than technology, is being adapted or reinvented across local contexts as the response to climate change advances.

Third, and in the long run most significant, the new centrality of place and the governance changes that accompany it put pressure on the organization of democracy itself. Sometimes—but rarely, we expect—the shift to the local will be narrow and self-limiting, as in the case of renewables in California. Much more typically, we think, growing local concern with sustainability will lead to changes in governance that spark new forms of democratic engagement. In both Ireland and the Amazon, collaborative review led to new forms of participation—the green municipality pacts in the Amazon, and the LAWPRO process for determining how and in what order to tackle cleanup projects—that promise key actors a central role in decision-making. In both cases, moreover, actors are asked not only to express their preferences but also to collaborate in the production of complex public goods—extension services—customized to their needs, and contributing to their own capacity to develop further. The challenge for decarbonization in both places, of course, is to make these reforms succeed. Progress at the frontier of decarbonization depends on innovations that change the way citizens make themselves heard in government—and the nature of the goods they can expect from it.

This last conclusion is necessarily tentative. But it would be as misleading to overlook the new forms of participation overflowing the banks of the familiar model of local democracy as it would be to claim that the flood had already cut and settled into a new course. One consequence of the new localism is already clear: the growing importance of local decision-making—including action at the level of sectors as well as places—diminishes the importance of action at the level of the entire globe. The story of the Amazon shows that even the biggest problems have to be solved close to the ground. In the next chapter, we ask what role this vision leaves for the global climate regime we have inherited.

6

International Cooperation beyond Paris

The easy part of imagining a global response to climate change is picturing the trajectory of jurisdictions that are already highly motivated to act. Most of Europe, the politically blue coasts of the United States, and a few countries—such as Japan, New Zealand, and Uruguay—have committed to industrial and agricultural policies that are broadly consistent with deep decarbonization. And at the federal level, at least for now, the United States has made climate change a central policy priority.

The trouble is that these efforts span only a small part of the global problem. Most of what these jurisdictions are doing remains focused on home markets, and only about a fifth of global emissions today come from places that are investing heavily to push the technological frontier, and apply the latest technologies and practices.[1] There are several reasons why we must fill this gap with a robust new vision of international cooperation—one that makes good on the promises of experimentalist governance.

First, international cooperation can accelerate and extend the reach of innovation. At the knowledge frontier, cooperation broadens the scope of parallel search. When Swedish policy makers and the nation's steel industry commit massive resources to experiment with essentially zero-emissions steel production technology, they can at best test one idea. By itself, Sweden's steel industry isn't big enough to invest in a more diverse portfolio.[2] Teaming up with others—in Austria, France, the Netherlands, the United States, and Japan—can expand the zone of exploration by testing other methods for decarbonizing steel.[3] Nobody knows which approach to cleaning up steel production will be most viable, but cooperation can ensure a broader range of experimentation. International cooperation can also create bigger markets for innovative products, such as by setting standards for products like aircraft that trade and operate globally.

Second, when innovation takes the form of contextualizing technologies for local markets—as California and other places are doing with renewable power (chapter 5)—cooperation helps other jurisdictions explore and learn about their own place-based solutions faster. Thus grid operators in Australia, California, Germany, and Ireland can (and do) compare strategies for integrating renewables; each faces a common problem and common points of comparison, even as each must find its own local solutions.

Third, cooperation is vital even in areas where solutions do not depend on technological innovations from advanced countries. Take deforestation. As we saw in the last chapter, stopping illegal logging and the burning of forest lands by squatters in the Amazon or Indonesia requires new, local practices, especially support services to small landholders who might otherwise encroach on forests. But these practices are also nested in national and international standards, backed by a penalty default: the threat of exclusion from markets where forest-related products are sold. International cooperation therefore helps make local reform work.

Fourth, international cooperation is essential for maintaining action in regions that are already doing the most. Innovators will check their efforts if they must compete with firms allowed to avoid the costs of decarbonization, and the prospect of such free riding weakens the political pressure for technological experimentation. Consider how this is playing out in Europe. Even in the greenest countries with the longest track records of innovation, governments regularly insulate those industries most exposed to global trade from especially burdensome regulatory requirements.[4] Sweden, for example, is pushing for cleaner steel production even as the government mostly exempts its domestic steel industry from the nation's stiff carbon tax so it can compete abroad.[5] This is where international cooperation—across the whole global steel market—has a crucial role to play. Until innovation makes cleaner steel production viable and strict rules in all markets make it commercially competitive, Swedish policy makers won't much disadvantage their national champions. As countries adopt costly boundary-pushing policies at home, industries press for trade policies—such as border carbon adjustments—to help level the playing field for cleaner but higher-cost European producers.[6]

For all of these reasons, no response to climate change will succeed without international cooperation. In this chapter, we explore how to achieve more cooperation of the kinds needed to accelerate decarbonization.

We are mindful that the quest for cooperation comes at a time of pervasive uncertainty in the world order. To the uncertainty already inherent in problems like climate change now is added uncertainty about how cooperation is to be achieved in today's international system—and even whether, in crucial areas, cooperation is possible at all. Yet despite these real obstacles, our story is a positive one. The practical need to learn from others, often combined with political coalition building, is already giving rise to a new climate change regime—one that looks to the Paris Agreement and other such international accords to provide legitimacy for experimental and locally engaged efforts and to create the penalty defaults that help give them force. But the real work occurs far beyond the strictures and procedures of Paris, in sectoral organizations and specialized federations of localities.

To see why, we must start by considering the state of cooperation today.

The Current State of International Cooperation

International climate cooperation does not begin with a blank slate, of course. As we described in chapter 2, there is a long international history of institution building for climate cooperation, most of it under the aegis of the UNFCCC. Formal failure in Copenhagen, long after the de facto failure with the Kyoto Protocol, opened up space for new ideas, which took preliminary form in the Paris Agreement of 2015.

In parallel with UNFCCC diplomacy, many other arrangements have been created for or repurposed to climate change functions. These include funding programs through multilateral development banks (e.g., the Global Environment Facility), new funding mechanisms (e.g., the Green Climate Fund), and institutions centered on particular regions (e.g., the Arctic or Amazon), sectors (e.g., the International Civil Aviation Organization), and pollutants (e.g., the Climate and Clean Air Coalition, which concentrates on soot and methane). Before Paris, governments pledged to direct at least $100 billion annually in new climate-related funds to developing countries, and the Paris Agreement extended that political commitment through 2025. At a meeting in Glasgow, Scotland in late 2021, donor countries were still falling short so they made new diplomatic commitments to spend more and check progress. Outside the ambit of the UNFCCC, there are also established regimes in forestry, palm oil, and other agricultural commodities as well as many nascent international regimes that coordinate actions in industrial sectors, such as cement, steel, renewable power, and oil and gas—all industries where a critical mass of firms has learned that it must collaborate on climate change.[7]

In short, a regime complex of partially cooperating, partially competing organizations has emerged. Its components are so numerous and varied, however, that it is a matter of intense debate which of them actually help solve joint problems and protect the planet.[8] The challenge for international cooperation on climate change isn't about creating new institutions on a blank slate so much as identifying and coordinating the efforts of those that do or could work.

This is where the Paris Agreement enters the picture—or at least where it was supposed to. The expectation (including, for a time, ours) has been that Paris was destined to become the central node in this regime complex, loosely coordinating both national programs and international organizations.[9] (Some saw a bigger and more directive role for Paris—as the master orchestrator, formally or de facto, of action dispersed across many organizations.) The reality, as we see it, is that Paris is and for the foreseeable future will remain less central than hoped. What little orchestration and coordination it does is of loosely coordinated clusters of efforts.

How are we to explain this situation? The adoption of the Paris Agreement in 2015 was met with widespread enthusiasm, including by us.[10] That support reflects, in part, success in abandoning elements of earlier failed efforts at cooperation outlined in chapter 2, such as the setting of global, binding emission targets and timetables for every country.[11] Compared with Kyoto, Paris adopts a much more flexible approach, based on national pledges known as nationally determined contributions (NDCs). These NDCs, the result of domestic politics, are to be restated every five years, reviewed individually through a "facilitative multilateral consultative process," and assessed collectively through a "global stocktaking."[12]

In principle, this structure has several things to recommend it. It creates a process that could align commitments with what national governments think they can implement.[13] (Parallel processes aim to elicit pledges from businesses, subnational entities, and NGOs—although there is no organized review and assessment of those actions that has much impact on behavior.)[14] It appears to dispense with a bright-line distinction between industrialized and developing countries by allowing countries to determine what they want to commit. It also includes large commitments for new funding—a critical element of the success of Montreal. Relinquishing old Kyoto-like efforts to create integrated global agreements was a big accomplishment, and Paris marked an end to those failed efforts. On the surface, then, many elements of the Paris Agreement appear to be aligned with the forms of experimentalist governance that we have been advocating. Indeed, in many ways Paris, especially at the start, seemed reminiscent of the experimentalist approach of the Montreal Protocol in its reliance on the periodic revision of national goals, review of ongoing efforts, and stocktaking of national plans and results.[15]

But in contrast to Montreal, Paris's initial framework has not been allowed to evolve, formally, toward experimentalism. Indeed in important ways it has not been allowed to evolve at all. National governments, protective of their autonomy, quickly recast Paris's initially permissive institutional rules into reporting and review procedures before any practical experience with them had been gained.[16] These procedures, known as the "rulebook," were hammered out through painstaking diplomacy that mostly ended in 2018.[17] The rulebook determines the minimum content for NDCs and emission inventories as well as the maximum granularity of reviews of NDCs and inventories, facilitative multilateral consultative process, and stocktaking. Fear of potentially intrusive processes kept reporting and reviewing mechanisms weak and general so as not to endanger consensus.[18]

In our view, these features of the Paris rulebook make it impossible for the machinery of Paris to serve most of the major functions of experimentalism. If NDCs are thin to begin with, no system of review can supply the missing content, let alone render assessments compelling enough to go beyond encouragement or chiding.[19] Take the case of Norway, a country highly committed to combating climate change and highly supportive of the Paris process. Indeed, adjusted for its size, Norway probably invests more in serious climate change policy than any other nation on the planet. The country issued its second NDC in 2020, strictly following the rulebook guidance. The document is almost completely procedural. Its sixteen pages are silent on the topics where Norway has taken the biggest risks and learned the most—such as efforts to push the adoption of electric vehicles, massive programs to advance carbon capture and storage among other pivotal technologies, and a pathbreaking effort to fund the protection of tropical forests, starting in Brazil. Its NDC likewise says nothing about balancing international engagements with the cost of efforts at home—a crucial consideration for a country like Norway, which is already so clean that reduction of domestic pollution is at the point of diminishing returns.[20] This reticence does not result from fear of disclosure. The country has nothing to lose from a thorough discussion of its plans; in several other forums, Norway has been highly transparent about its climate changes initiatives.[21] If a committed actor and eager learner like Norway sees no purpose in engaging deeply with the NDC process, what should we expect from those less committed to the industrial transformation needed to reduce emissions?

The same consensus-driven orientation that plagued the rulebook also impedes exchanges with nongovernmental and subnational actors—among the most important sources of political pressure and practical action on climate change. In the run-up to Paris, many governments wanted to acknowledge explicitly the potential contributions of firms and subnational jurisdictions to cutting emissions. In response to their concerns, a registry—the Global Climate Action Portal—was developed, in which nearly thirty thousand "actions" by more than eighteen thousand subnational actors have been declared.[22] But the climate change secretariat that manages Global Climate Action has no authority or capacity to help registrants advance their projects and coordinate with others. Even when the secretariat orchestrates reviews of the NDCs, following the procedures set forth in the rulebook, it is explicitly barred from looking at the Global Climate Action Portal information unless sovereign governments themselves include those actions in their national plans.[23]

There are other limits of these kinds as well that make the break with the past less clean than first appeared and constrain the promising articulation of promising initiatives. While the Paris Agreement does away, to some degree, with the categorical distinction between developed and developing economies, vestiges of the division persist.[24] Paris has created some technical cooperation mechanisms, like a series of Technical Expert Meetings, that look, on the surface, like the experimentalist bodies of the Montreal Protocol.[25] But these bodies—organized by a secretariat that was expert in negotiations, not the frontier testing of new technological ideas—have achieved little because they have neither the authority nor capacity to organize the consequential searches for information that characterize Montreal. Mostly they rehash what is already known.[26]

As for large pledges for new funding, Paris imposes weak, uneven controls and accountability on the exact level of actual contributions—and no mechanism for countries to report what they have attempted to achieve with the funding, where they have fallen short, and where they have learned they can do more.[27] Such mechanisms have been left to other institutions to develop and for national governments to innovate, as they please, on their own. Unlike Montreal, moreover, little of the funding mobilized with reference to Paris has been (or will be) spent through institutions formally governed by Paris.

For all of these reasons, we no longer think it is likely that governments could demand highly informative NDCs, extensive peer review, and detailed stocktaking under Paris. However appealing the vision, the rulebook is the reality; Paris's consensus procedures will not give coherent direction to deep cooperation needed to manage climate change. But that does not mean that Paris has no role to play in serious efforts at cooperation. Indeed, international coordination must leverage the legitimacy of the Paris institution and its role as the de facto climate conscience of the global community, with its corresponding capacity to punish parties that drag their feet.[28] To make the most of Paris means relying on the agreement as a source of authority and penalty defaults while performing nearly all the work of experimentalism on the outside.

The central challenge for governments, firms, and NGOs seeking to advance effective cooperation on climate change is thus to sustain the legitimacy of Paris while not becoming mired in the machinery. Achieving that will not be easy—in the eyes of many, the machinery is the legitimacy—and the rest of this chapter offers a framework for how to manage the challenges. We focus on the central, experimentalist functions needed for effective cooperation to be performed.

Our discussion is based around two poles of industrial organization. At one extreme are globalized industries that both source and sell in world markets; examples include industries for aircraft, solar panels, aluminum, petroleum, and oil-drilling and production equipment. At the other extreme are industries that sell intrinsically place-based products, such as fruits or vegetables characteristic of particular regions, residential homes, or local electric supply.[29] At the former extreme, most change comes from frontier innovation and penalty defaults linked to international standards; the role for cooperation revolves around standards setting and the application of penalty defaults. At the latter extreme, contextualization—local building codes, say, or seed varieties optimized to particular soils and climates—plays a bigger role, even when local production relies on technologies that are sourced in global markets, such as heat pumps to increase energy efficiency, precision irrigation systems to economize on water, or advanced solar cells and wind turbines. Here the gains to cooperation come largely from accelerated local learning—motivated by penalty defaults eventually in combination with international standards that level the competitive playing field.[30]

We consider these two cases in turn. First, extending the arguments in chapter 4, we look at international standards-setting actions that can push the technological frontier. Second, we explore the contextualization of technologies, building on the arguments in chapter 5. Third and last, we examine the penalty default incentives that intersect with both of these efforts—and the important role Paris has to play in them.

Cooperation and the Technological Frontier

We begin with international standards setting, where the incentives for cooperation are the strongest and the lessons are the clearest. By our estimate, about half of the world's emissions are directly subject to international standards linked to rules of market access. That includes almost all emissions from industry (21 percent of the global total emissions), most emissions from transportation (24 percent), and about half of food and land use emissions (24 percent of the world total).[31]

Deep decarbonization in these markets requires deep innovation. As a measure of the level of the challenge, a recent study by the International Energy Agency found that up to three-quarters of the technologies needed for deep decarbonization do not exist today, and in every major cluster of technological innovation—except for solar power—the global frontier for the testing and deployment new technology is far from where it must be if global emissions are to be cut dramatically to zero by midcentury.[32] Here cooperation has an essential role to play. As we saw in the preceding chapters, regulators use technology-forcing requirements to induce firms and research organizations to exchange information about what is becoming feasible. And because their exchanges percolate into national and international standards—for example, CARB's vehicular emissions goals initially shape the EPA's national standards, and then standards elsewhere, especially those in the European Union and Japan—the prospect that local standards will spread motivates other firms and jurisdictions to try to innovate, or at least keep pace with innovation, lest they be left out.

Sector-based innovation is driven of course by firms and governments motivated to push the frontier. Those efforts, as we have seen, benefit from regular stocktaking to tighten or loosen goals and help identify opportunities for new investment.[33] As we noted, international cooperation to encourage frontier innovation faces two closely related challenges.

The first and most central challenge is striking the right balance of commitment with openness. The goal is to build a coalition of governments and firms all motivated to invest in experiments, review results, and codify standards while remaining open to new participants from outside the circle of innovators. These coalitions might be what the literature in economics and political science calls a "club": a group that provides a collective good, but only to members.[34] Or they might be looser federations of like-minded actors pursuing broadly similar goals. (We think that there is a large role for clubs in advancing climate change action—a topic that we revisit below with our discussion in the next chapter of trade measures.) The challenge is to begin with a coalition small enough that the incentives to innovate and the insights are focused yet prospectively broad enough to secure a measure of legitimacy. Nearly all existing sectoral institutions have large memberships and decision rules that together can easily replicate the consensus gridlock that plagues the UNFCCC and Paris.[35]

New approaches—which allow small groups to start innovating while keeping their membership open—are beginning to play out in aviation and shipping. In both sectors, organizations with global membership (such as the International Civil Aviation Organization and International Maritime Organization) are being pushed from within by smaller clubs of firms and governments mainly based in Europe.[36] These small groups recognize that the global adoption of more demanding standards depends on demonstrating the technical and economic feasibility of new, more effective technologies and methods. Thus Maersk, the world's largest container shipping company by fleet size and cargo volume, coordinates a series of technology demonstration programs inside the International Maritime Organization cofunded by governments and linked to proposals for new standards.[37] Those demonstrations are evaluating different fuels such as natural gas, biofuels, hydrogen, and ammonia in combination with internal-combustion and electric-drivetrain propulsion systems. Because Maersk's capital stock is long-lived and hard to change once built—onboard systems and shore-based infrastructures are codependent—the company also works with those same governments to gradually align equipment and local standards to superior solutions. This process shows the workability of many paths to improvement and makes it easier, eventually, for other International Maritime Organization members to join in and riskier for members to drag their feet.

There is thus a tension between a compact vanguard and a rearguard whose inclusion ultimately determines the legitimacy of the overall effort. And as we will see in the next chapter, this tension poses a generic challenge in international cooperation—especially given the decline of hegemonic leadership and deference to technocracy. In many stalemated domains of world politics, including trade and the WTO, innovative cooperation has shifted to smaller groups operating at the margins of consensus rules and constantly aware of the need to make new norms acceptable to outsiders threatened by them. A vanguard therefore pulls a rearguard along, and in time, the two perhaps merge.[38] In climate change, where new technologies are costly, and require testing and nurturing in protected markets, the strains of maintaining broad consensus while establishing the autonomy to experiment—and then making the results of innovation broadly accessible—are particularly severe.

The second major challenge for more innovation-oriented cooperation is granularity. What is the right scope for a "sector"? Experience leads in contradictory directions.

On the one hand, many examples suggest that the narrower the focus, the greater the progress. In the North Sea, pollution from incineration was handled distinctly from ship dumping and agricultural runoff. In the Montreal Protocol, fire extinguishers and other halon users were evaluated separately from the use of CFCs, carbon tetrachloride in dry cleaning, and methyl bromide in fumigation. Even within sectors, the focus could shift under Montreal as the nature of "the" problem subtly changed; the first generation of CFC refrigerants were phased out on a different schedule from the second and third generation, for instance, with distinct clusters of experimental innovations that pushed the frontier in each of these chemical classes.[39] And as we saw in chapter 4, when utilities and regulators first sought to control sulfur pollution, they adopted distinct approaches to experimentation around each control option—linking the efforts together into a more nearly industry-wide program only after the potentials for progress on each narrow frontier were better understood. Underlying this drive to ever-sharper focus and narrower sectoral definitions is the need to compare plausible alternatives in peer review. That entails fine-grained comparisons of rival solutions for a known commercial function (e.g., low-emission yet highly reliable jet engines) and setting standards for tangible products that actually move across borders.

On the other hand, fine-grained approaches may not always be best. Under conditions of technological uncertainty, where solutions are often derived from innovations in other, unrelated sectors, it is costly to design institutions for closure and narrowness; instead, an openness to developments beyond the usual boundaries is frequently indispensable, even as solutions to particular problems take shape. A fine-grained sectoral group will have neither the experience nor the authority for open-ended exploration. Indeed, within the Montreal Protocol the technological options committees were sector based but not *rigorously* sectoral; they could shift their scope as needed to address a cluster of related ODS. For example, efforts to phase out hydrochlorofluorocarbons (one of the first substitutes for CFCs) and hydrofluorocarbons (a later substitute that is more benign for the ozone layer, but still harmful for global warming) have coevolved. The regulation of hydrochlorofluorocarbons depends in part on the options for switching to hydrofluorocarbons, and that in turn depends on the next tranche of superior alternatives along with the equipment to use them.[40]

Similarly, in the area of zero-emission steel, nascent efforts to organize the sector have identified a small array of possible solutions: some involve electricity, others involve carbon capture and storage technologies, and still others involve the use of hydrogen in steel production. The steel industry and supporting governments can organize experiments, review results, assure procurement and protection for early movers, and set standards for the industry. But success of the enterprise also requires actions outside the fence line of the steel industry and its regulators. Electric methods for making green steel require clean electricity, carbon capture and storage approaches would benefit from prior demonstrations of industrial carbon capture and storage systems, and hydrogen-based strategies hinge on adequate infrastructure for making and supplying hydrogen at a reasonable cost.[41]

The granularity problem underscores why strategies for innovation and standards setting must, following this example, be sector based but not sector bound: anchored in the concrete concerns of a certain sector, but not by the boundaries of the sector in the search for solutions. One model of this kind of directed but open-ended innovation is ARPA-E, as we discussed in chapter 4. The agency specializes in solving hard problems that block progress in a given sector by looking simultaneously inside and outside for solutions. Its early stage project funding connects ideas *across* sectors to open up new possibilities. Better technologies for bulk energy storage, for instance, require thinking about the frontier for innovation not as "energy storage" but rather as combinations of skills drawn from areas such as nanomaterials and chemical engineering.[42] A similar sector-based logic applied when EPRI helped the power industry learn about scrubber innovations. To ensure widespread deployment, EPRI, its member utilities, and the industry's equipment vendors had to look beyond the performance of different chemical engineering systems to the supply chains for critical materials (e.g., limestone) and disposal.

While most of these examples come from *within* countries, the same logic applies to international efforts. Here, the EU experience with hydrogen is emblematic. Most studies find that deep decarbonization is best achieved by converting as many energy uses to electricity as possible and then decarbonizing the electric power system.[43] But that approach faces extreme difficulty in a few sectors for which electric solutions seem unlikely to appear, such as the manufacture of low-emission plastics, steel, and aircraft.[44] In these cases, and possibly many others, hydrogen could be the key to eliminating emissions. The crucial unknown on which the entire strategy turns, of course, is the cost of making hydrogen. The European Union has organized a multinational effort to answer that question. It includes ARPA-E-style funding for early stage research and also (unlike ARPA-E) direct funding for promising experiments at scale. Aggressive peer review at each stage is keeping tabs on which production routes seem most promising, while also keeping an eye on the political backing needed to sustain the program.[45] Within international organizations, this kind of sectoral approach requires navigation around the inevitable blocks that arise when the interests of a particular member change. The bigger the group and the more consensus-oriented the decision authority, the harder that navigation will become.[46]

At this writing, a lot of boundary pushing is underway in nearly every major emitting sector.[47] Nearly all of this investment is occurring with reference to Paris—in the name of Paris—yet none of it is organized "under" the Paris Agreement.[48] It is happening in small coalitions of governments and firms focused on key challenges, such as improving the performance of light-duty electric vehicles, building better nuclear reactors, or lowering the cost of hydrogen production. The configuration of each group varies with the task at hand. The best performers operate in experimentalist modes with active peer review, evaluation of results, and adjustment of central rules—all motivated by penalty defaults.[49]

Cooperation and Contextualization

We have argued that sector-based international standards can have direct, powerful impacts on the direction of technological change. But the reach of sectoral standards is often limited, especially when technologies, products, or investments don't move much across borders. Even where firms and governments are under intense pressure to act on climate change—as evident in the many cities and regions that are now making bold policy announcements in forums like C40, ICLEI, and "we are still in"—what happens locally depends on local factors revealed through contextualization, not international standard setting. When the electric utilities in California were under pressure to integrate more renewables, for example, they didn't look to the International Renewable Energy Association, a global body, for help in solving problems. Rather, they consulted other state utility regulators, grid operators, and politicians and aligned their own investments and operations with local expectations. The local context determines almost everything about how these technologies get applied. International organizations can supply data—for instance, how California's efforts compare with others—and suggest how California's track record might be applied in other localities keen to replicate California's successes.

The role for international cooperation in directing and accelerating contextualization will thus be a lot messier than in the highly tradable sectors discussed in the previous section. The incentives for joint action on contextualization are more diffuse. Innovators are not automatically compelled, through foreign competition, to drive international standards higher or align their operations with those standards. Instead, many of the gains from cooperation will come from joint learning from the successes and failures of contextualization strategies. Ideas developed elsewhere can seldom be transferred unchanged from one place to another. Effective learning requires an active effort. And almost invariably, such learning must look beyond the exchange of best practices to joint efforts to understand, through joint review and other means, how solutions are reconceived as they are transferred.

Fortunately, there is a rapidly expanding set of institutions that may provide models for this kind of review. One example is the National Association of Regulatory Utility Commissioners (NARUC) in the United States. NARUC provides detailed peer review and the sharing of best practices on electric utility regulation. Through committees on topics from rate design to the reliability of natural gas supply and recovery of investments in innovation, NARUC helps North American regulators learn how other North American regulators have innovated—what has worked, what hasn't, and why. These reviews help set local agendas for contextualization. When the US National Academy of Sciences outlined a vision for how the electric grid can be kept reliable and affordable while addressing new societal goals such as deep decarbonization, it pointed to NARUC in particular as one of the critical institutions that will need to help each state contextualize the lessons being learned in other venues around the nation.[50]

Increasingly, NARUC plays similar roles overseas. It collaborates with foreign regulators, themselves also often organized into collectives for assessing experience with contextualization. Such assessments begin by identifying a frontier topic in electricity regulation and establishing direct partnerships with regulatory counterparts to investigate it together. Where NARUC collaborates with peers that are also grappling with the same issues—notably in

Europe—the information flows are in two directions. Where NARUC operates more to build capacity overseas, such as in emerging and least developed economies, the information flows are more unidirectional. NARUC members also often fund direct capacity building—authorizing that effort through institutions formally outside NARUC such as the Regulatory Assistance Project that specialize in the regulation of monopoly industries.[51] Through this kind of detailed spadework, the lessons learned through contextualization, frequently at great local cost, spread more rapidly and widely overseas.

While NARUC runs peer reviews within place-based sectors, a similar logic applies to certain cross-border problems such as the peer review of national North Sea pollution. The Netherlands led the other littoral nations in taking the North Sea pollution problem seriously and creating effective peer review by volunteering to go first.[52] Similarly, the United States and China volunteered themselves for the peer review of their policies on fossil fuel subsidies as a way of laying a foundation for mutual cooperation on energy and climate.[53] Each government appointed its own experts, who along with other internationally recognized peers, published a mutual review of subsidy policies and identified the need for further reforms.[54]

Most contextualization, to repeat, occurs *within* countries—often far from capitals and highly localized. It is tempting to avoid the complexity and cost of these processes by relying on centralized interventions, including foreign pressure and support, to secure the desired local outcomes. These efforts reflect the triumph of hope or wishful thinking over experience, as attempts to control uncertain situations from afar repeatedly fail. Stalled efforts to create a viable program for protecting the Amazon in Brazil are an object lesson of the high costs of such apparent shortcuts.

At the same time, as we have emphasized throughout, local direction is hardly sufficient. The experience of many US municipalities that have recently committed to ambitious programs of decarbonization is one of many illustrations. They have announced bold goals and built alliances to pursue this end.[55] But for the most part, local capacities remain modest, and efforts to pool experience are likewise in their infancy. Ambitions to greatly expand joint learning are robust, but the capabilities to do so are limited—precisely because most of the pioneer jurisdictions have not framed the climate change problem in this way. Many states and cities have sustainability offices that have taken on the climate change agenda, but these institutions don't have the capacity to contextualize new technologies and best practices. Many have set goals and then struggled to figure out how to meet them.[56] In some domains, local city or county actors often have no authority to alter the practices that cause emissions—for example, the industrial production of imported products or the configuration of the electric grid. In other domains, the leverage is greater—for instance, regional mass transit planning, infrastructure for electrifying vehicles, and building codes. In the absence of decisive action on climate change at the federal level, it seems that municipalities and states in the United States, like those in Brazil, need to develop their own solutions where they can, learning from each other, and eventually using what they learn together to leverage action at the federal and international levels.

We know it is much easier to say all this about contextualization than to do it—and in fact, it too often is said rather than done. Above all, we wish to emphasize that contextualization requires local learning. Institutions for international cooperation can be helpful only to the extent that they focus local efforts by sharing information and experience across jurisdictions.

This is where the current efforts can stand to be improved. Subnational actors that have invested the most in figuring out how to put emission controls into their local context have tried to learn from efforts elsewhere by forming several associations. C40 and ICLEI are, at this writing, the most prominent—both large clubs of subnational units that have pledged actions on their own, typically in line with deep cuts in emissions as needed to meet the Paris goals. They are expressions of passion and desire. But neither operates a serious peer review program, and neither administers a capacity-building program—either through multilateral funding akin to what is offered under the Montreal Protocol or by direct regulator-to-regulator collaboration along the lines of NARUC.[57]

Cooperation and Penalty Defaults

We began this chapter by emphasizing why international climate cooperation must look beyond Paris. But in our view, Paris still has an indispensable role to play, albeit in an area that Paris participants, bound by the founding commitment to national sovereignty, are reluctant to discuss: penalty defaults. Paris is essential to allowing the imposition of penalties—more exactly, to marshaling support for the kind of penalty defaults without which experimentalist regimes won't work.

Climate change diplomacy has been dogged by the question of whether to enforce compliance by penalties, and if so, how. In one camp are those scholars who view climate change as a giant collective action problem—a prisoners' dilemma on a global scale that will thwart collective action unless obligations can be comprehensively agreed on and enforced. In the other camp are those diplomatic pragmatists who see agreement on enforcement as impractical, along with those academic theorists who see many problems of "compliance" as resulting not from willful violations but instead from incapacity—for which the best response is support for capacity building rather than penalties.[58]

Experimentalist governance splits the difference. It sees penalty defaults as necessary to undermine confidence in the stability of the status quo—thus encouraging hesitant innovators and reluctant compliers—and raise the cost of persistent obstinacy or indifference. But experimentalism stresses that countries strapped for resources will frequently struggle to master tasks that challenge rich countries as well; that noncompliance will therefore often result from incapacity, not selfish guile; and that the response in such cases should be support rather than sanctions.

Where does Paris fit into all this? On the one hand, the Paris Agreement lacks the formal power to establish penalty defaults because mitigation goals are set by the various parties in their respective NDCs. In other words, Paris does not have goals of its own that are applicable to individual members, and cannot penalize parties for ignoring requirements it has not imposed. The agreement's ambitious collective goals—stopping warming well below 2°C above preindustrial levels and achieving roughly net-zero emissions by midcentury—were acceptable precisely because they were collective; they did not make particular demands of particular parties. But on the other hand, Paris enjoys unique legitimacy among international organizations as the authoritative voice of public opinion in global climate affairs. This legitimacy is rooted in the very legacy of consensus decision-making that otherwise hamstrings the Paris process in so many ways. Its role in threats to climate is like that of the United Nations in threats to peace: whatever the shortcomings of the institution, there is practically nowhere else to turn in a crisis.

This arrangement explains why appeals to Paris are made when a government or firm is seen as willfully undermining decarbonization. That was the reaction when Brazil all but boasted of invading the Amazon after years of varying efforts to protect the forest; almost instantly, European governments threatened trade retaliation against a flagrant violation of the goals of Paris. For similar reasons, the European Union treats membership in Paris as a kind of prerequisite to entry into the fellowship of responsible nations and has declared that it will not enter into trade agreements with nonmembers.[59] EU members have also suggested they should sanction the United States for its failure to honor Paris under Trump.[60] The Biden administration's plan for climate action, which envisions much stronger policies at home, has at times included broad measures and sanctions to push the rest of the global economy along too—all in the name of advancing the Paris Agreement.[61] For their part, NGOs with decades of experience shaming corporations into climate action, from Friends of the Earth to Extinction Rebellion, now refer to Paris to help legitimate their demands and campaign for Paris itself to act more boldly. In both efforts, they acknowledge Paris as the climate conscience of the world.[62]

Trade measures are all but certain to play an important and growing role in inducing action as well as contributing to the legitimacy of Paris and efficacy of climate governance.[63] Sanctions, actual and threatened, make the stewards of globalization, especially trade lawyers, nervous. What will keep sanctions and other types of penalty defaults focused on appropriate goals (boosting the legitimacy of Paris and mitigating climate change) while preventing them from spinning out of control into instruments of protectionism and geopolitics? One answer is that sanctions will have to respect the existing trade norms, which require nondiscriminatory application that advances legitimate environmental goals, including goals established through international cooperation.[64] The precedents that gave rise to these norms will constrain the use of trade measures and make threats to apply them more credible.

Nonetheless, the current use of penalty defaults is unstable, and from the point of view of experimentalist governance, incomplete. Instability in the use of penalty defaults arises from the ease with which accusations of noncompliance and threats of sanctions can be made, and the lack of any structure within Paris for responding to those claims. When serious threats are easy to make and hard to challenge they will be made too often, draining them of meaning.

The current arrangements for the use of penalty defaults are incomplete, at least from the perspective of experimentalist efforts, because the growing risk of a violation does not trigger an offer of technical support to help facilitate compliance. As we argued in chapter 3, under uncertainty, where learning will typically be necessary to meet changing requirements and yet those learning capabilities are costly for many nations to obtain, a complete system of incentives for encouraging compliance must include not only penalties for incorrigible laggards but also support for those who do or could lag for want of help in mastering new tasks. Yet the funds linked to Paris today include only modest capacity building. Moreover, because the

Paris rulebook prohibits a close look at country performance beyond the information provided by countries in their NDCs and regulatory transparency reports, there is no formal way to target more capacity building to those parties whose performance shows they need help.[65]

A modest reform can begin to address both of these concerns. In cases where incapacity is clearly the cause of poor performance, parties to Paris could agree to recognize noncompliance penalties only if they are accompanied by promises from the accusing countries to provide technical and financial support. To demonstrate the need for support, countries charged with noncompliance could volunteer for a peer review with an outside partner that would establish whether incapacity caused the shortfall, and if so, indicate the first steps toward a remedy. Parties that egregiously undermine Paris goals, like Brazil, would never put their misdeeds on public display in a peer review; rich countries would be reluctant to participate too, either for pride or fear that a thorough review would expose politically explosive lapses and delays. Instead, the beneficiaries of such a reform would be the intended ones: developing countries most likely to need support and make good use of it, along with those most likely to reject financial obligations for Paris as a whole. With this change—pairing accusations of noncompliance with promises of help where it is needed—threats would be costly to make and thus more likely to be made judiciously. The incentives for changing behavior would be more complete and credible than the episodic menace of penalty defaults alone. This kind of conditional support could be financed from the $100 billion of new annual assistance confirmed under the Paris Agreement, and if there are pledges of funds beyond the $100 billion, this would be a good way to spend those additional resources. A mechanism quite similar in approach—financial and technical support, in exchange for peer review that roots out the causes of noncompliance—has improved technical capabilities and compliance in the area of customs modernization, as we discuss in chapter 7. It shares many features with the MLF of the Montreal Protocol, with monetary policy reform supported under the International Monetary Fund and in many other domains where the ability to cooperate and comply must be built within countries.[66]

The legal authority for such an approach already exists, broadly, in the Paris Agreement.[67] The European Union is in a good position to lead the way here, for it is already threatening trade measures against countries that make inadequate efforts. That kind of penalty default could be coupled with incentives for countries that are willing to explore needed changes in national policy.

The Future of Cooperation: Beyond Paris, with Paris

Most progress on international climate cooperation, we have argued, will come through standards setting at the frontier of technological innovation and, to a lesser extent, contextualization that pools local learning. The roles for Paris in performing these functions is small yet vital; it can serve as a point of focus and conscience that helps make each of these three major functions more effective. The jurisdictions that are most motivated to act on climate change know that making progress requires international collective action. By putting an end to Kyoto-style diplomacy and declaring the beginning of an alternative order, Paris became the constitutional foundation of that global cooperation. But in interpreting Paris's role, it is useful to keep in mind journalist Walter Bagehot's distinction between the "dignified" part of a constitution, whose function is "to excite and preserve the reverence of the population," and the "efficient" part, which serves to "employ that homage in the work of government."[68] No global consensus organization, even one as flexible in its system of pledges as the Paris Agreement, will ever be the "efficient" part of an effective integrated system for experimentation and rapid learning. Yet the same need for a continual renewal of consensus that disqualifies Paris in this sense equips it uniquely as the dignitary face of global climate cooperation: the climate conscience of the world.

Paris therefore should be celebrated for what it does well: establishing the legitimacy and foci for climate action. The efficient work of international cooperation—pushing the technological frontier in small clubs and investing heavily in contextualization—will be done in the name of Paris, but far removed from the grinding machinery of intergovernmental consensus. Practically, Paris is most important in concentrating effort by other actors—including nations that want to threaten trade sanctions—to apply the penalty defaults that help motivate experimentation. That role for Paris necessarily implicates other international economic institutions, notably the WTO, that are themselves fragile and searching for relevance in a world where institutions of the old globalization are deeply in question. In the next and final chapter, we look at how that search may be leading to a trade regime, built in pieces, that is more hospitable to international efforts to mitigate climate change, and more responsive to demands for transparency and democratic accountability in governance across many levels.

7

Piecing Together a More Accountable Globalization

This book has been concerned, above all, with the pressing threat of climate change. But the central themes we have explored throughout—from the failures of uniform, global solutions to the urgent need for more context-sensitive and accountable experimentalist policies—bear on a far broader range of problems. In this final chapter, we examine how international trade, facing many of the same challenges as climate change, is beginning to develop experimentalist responses as well. We conclude by calling for a new globalization, arrived at piecemeal, in which greater openness to cooperation in the fight against climate change goes hand in hand with more attention to the dislocating effects of trade.

The parallels between climate governance and trade governance are striking. We have argued that the real work of climate change is done in sectors; the same is increasingly true in trade, as we will see in a moment. Moreover, the top-down consensus decision-making that has stifled adaptation in climate change governance has slowed adaption in trade even more. In both domains, the failures of global diplomacy and presumptions of technocracy have left pressing problems unsolved, and bred public mistrust of global institutions. The Paris Agreement greatly loosened those constraints in climate, but it did not create anything approaching a complete, workable alternative. A similar—if less dramatic—movement to loosen the grip of consensus has been unfolding within the WTO as well, and in the WTO, this rejection of the old regime is going hand in hand with the construction of a new one.

At the same time, pressure for the more direct integration of trade and climate regimes is increasing. As commitments to decarbonization in the advanced countries become more and more ambitious, and public pressure for enactment grows, so too does concern that investments in green products and production technologies will be undercut by imports of dirty technology. Alongside episodic efforts to establish penalty defaults, energy-intensive sectors such as steel, cement, and fertilizer have been organizing a systematic demand for a trade regime that conditions market access on respect for the evolving green standards.

How this integration should occur, if at all, is an open question. At one extreme, the imposition of green standards could be little more than a pretext for maintaining the protection of vulnerable domestic producers against lower-cost foreign competitors. At the opposite extreme, an experimentalist process of exploratory standards setting could be adopted sector by sector, using the institutional capacities for continuous monitoring and self-revision needed for decarbonization to remedy the accountability as well as responsiveness problems that beset globalization today. In between these two poles might be a purpose-built trade regime that harks back to Kyoto, and—ignoring what we see as its failures—uses market mechanisms to recover from dirty foreign producers the difference between what they spend on pollution reduction and what they would have spent to meet the higher standards of the export market.[1]

Our vote, of course, is for the experimentalist solution: an integration of trade and climate change adapted to uncertainty, and accountable to a public rightly suspicious of technocratic elites. Linked in this way, a trade and climate regime contains the germ of an alternative globalization that takes to heart the failures of the current one.

Globalization and Its Discontents

Kyoto was an expression of the international interval—brief, as it turned out—between the fall of the Berlin Wall and the rise of China as a great power in which it seemed plausible that in matters of global scope, the world should be subject to a single set of binding rules, administered by experts and responsive to markets. The formation of the WTO in 1994 marked both the high point of this movement and the end point of the postwar economic order's progression from cautious avoidance of the beggar-thy-neighbor policies that deepened the Great Depression of the 1930s to aggressive support for global free trade.

Yet almost from the moment it was put in place, the WTO came under attack from those that feared the disruptive effects of expanded trade under technocratic direction. The most vociferous and sustained opposition came from the "losers" of globalization: working-class communities in advanced countries that collapsed as jobs were outsourced and cheap labor was integrated into global supply chains. In time, the frustration of these communities, and their disappointment with the elites and political parties, coalesced into a populist revolt that is reshaping politics in advanced democracies.

In fact, the technocrats did fail these communities. Experts were confident that the overall gains of globalization would outweigh the costs—and that the losers were best compensated by ex post facto redistribution, not inefficient protection of their jobs and markets, which would only reduce the surplus available to them. Confidence in what theoretically had to happen dulled curiosity about what actually did. And even as the decline of industrial communities became unmistakable, trade was manifestly not the only cause. Technological development played an important role too as routine jobs—often an important rung on job ladders in mass production factories—were automated away.[2] Though it is now clear that trade dislocations—particularly exposure to Chinese imports—had an independent and detrimental effect on industrial communities in the United States, for a time it was hard to know whether trade was much of a problem at all.[3] Either way, whether from overconfidence in their own claims or professional scruples about the ambiguities in the data, political and economic elites neglected the trade dislocations that decimated the towns and cities of the old industrial economy. Poverty and hopelessness exploded in these communities. Older, male workers were scourged by "deaths of despair" through drugs, alcohol, and suicide.[4] Historians of the future will have the grim task of understanding how this could happen in country after country whose elites vaunted their commitment to democracy and shared prosperity at home even as they hoped that global trade would benefit the world's poor.

An Inevitable Disaster: The Globalization Trilemma

On a deeper view, though, the devastations of globalization were not just an accident waiting to happen, given the inattention of the elites, but rather a catastrophe foreordained by the incompatibility of the goals of a technocratically managed commercial world order.

This is the thrust of economist Dani Rodrik's characterization of globalization as a trilemma, which says we cannot have democracy, globalization, and the nation-state all at once.[5] The trilemma starts with nation-states, governed more or less democratically. Recognizing that global markets, like all markets, require a regulator to specify the minimum characteristics of the products allowed into commerce and the processes by which they are made, the nation-states authorize the creation of the WTO. As the global regulator, the WTO can be either democratic, reflecting the plurality of values that historically condition commerce in the democratic nation-state (the equity of partners in an exchange, and a concern for the environment and other externalities), or technocratic, focused on minimizing market friction.

This spare framework is enough to demonstrate that if we pursue globalization, we must sacrifice either democracy or the nation-state. Suppose the global regulatory regime is centered in a technocratic body, established by treaty among states, but only remotely accountable to them. In this case, the nation-state persists, but democracy must defer to the technocratic decisions that foster globalization. This is the condition that approximates what we actually have. Rodrik calls it "hyperglobalization": the global market becomes almost as uniform as a domestic one and therefore cannot accommodate the plurality of values that legitimate exchange in different places.[6]

Alternatively, imagine that the global regulator is democratically responsive to a global polity. In this case, there is global trade and (global) democracy, but the nation-state as we know it disappears, for the global polity has supplanted the nations. If we choose to safeguard democracy in the nation-state, finally, we must sacrifice global trade as there is then no place for a global market regulator, and no workable market without one.

Given these inherent limits of globalization, Rodrik argues, the real choice we face is between the current technocratic regime—and the waves of antiglobalization it provokes—or a return to pre-WTO rules of trade, which prevented the cascades of protectionism that destroyed world trade in the 1930s, but otherwise left each nation-state largely free to regulate itself. On this view, it will often (but not always) be in the self-interest of a nation to reap the benefits of specialization that come with trade. But either way, it is the nation that decides—market by market, not as an all-or-nothing choice between joining a uniform, global trade order and economic isolation—when trade makes sense, and it is the nation, not the other countries of the world, that bears the consequences of its decisions.

Where does the international response to climate change fit in this scheme? Although self-interest generally leads nations to participate in trade, Rodrik asserts, free rider and other familiar collective action problems cut the other way in climate matters. Precisely because the atmosphere is a commons and nations are not directly exposed to the consequences of their actions, Rodrik contends, they will do nothing in the absence of an agreement that binds them all. Correcting the excesses of hyperglobalization would thus require the reconstruction of the trade regime, and if there is the global political will for it, the creation of a climate regime alongside it. For better or worse, addressing the problems of trade leaves the problem of climate change untouched.[7]

The Decline of Tariffs and the Rising Importance of Nontariff Measures

As we see it, this understanding of trade is incomplete, and this partiality obscures possibilities for reform that could increase the national control of trade agreements as well as put the authority and resources of the WTO behind the global response to climate change. Indeed, at the same time that the trade regime came under massive pressure from outsiders, it was being changed from within by its very negotiating successes, and reshaped by the changes in the organization of the form and nature of the regulation described in chapter 3. Together these changes are refocusing trade on sectoral concerns in ways that respond to the accountability problems highlighted by the globalization trilemma and that facilitate the institutionalization of experimentalist decarbonization regimes of the kind portrayed throughout this book.

Regulatory concerns traditionally play second fiddle to tariffs and fees as factors in trade negotiations. Their relative insignificance is reflected in their generic characterization as nontariff measures. Successive rounds of trade negotiations, beginning in the early postwar period, well before the formation of the WTO, and continuing under its auspices, however, have reduced tariffs to nuisance levels, which the WTO glosses as so low that the cost of collecting a tariff exceeds the revenues. Meanwhile, the speed of innovation along with the growing complexity of products and supply chains have heightened the risks of latent hazards, exposing the regulation of international commerce to a distinct form of the same kinds of pressures faced by domestic regulators. Trading partners have found ways to address these concerns, but their solutions fit poorly within the formal structures allowed by the WTO, and the tension between the growing need for accommodation and the rigidity of the response generates pressure for reform.

At a minimum, to address nontariff measures, partner countries have to agree to formally list these kinds of risks as requiring attention and update their lists as circumstances change. But beyond this agreement, monitoring trade requires mutual confidence that dangerously nonconforming products trigger appropriate alarms. The inner mechanisms of regulatory systems—their inspection routines, incident reporting systems, and recall alerts—must come to similar conclusions about similar situations, regardless of the differences in working methods. To achieve this kind of demanding, regulatory equivalence, trade partners regularly review one another's inspection practices or carry out joint reviews—the kind of peer reviews by which, as we saw in chapter 3, organizations learn from their differences. For example, in a particularly well-developed variant—the 2011 Bilateral Air Safety Agreement between the European Union and United States—each regulator can certify conformity to design requirements or maintenance standards on behalf of the other. This experience is the basis for mutual influence in standard setting, yet each party, after exhausting a complaint and review mechanism, can withdraw approval (of a part) of a standard with whose development it has come to disagree.[8]

Such peer review regimes have emerged in areas like food safety, civil aviation, and pharmaceuticals, where the concern for hazards is most acute. They have also arisen between advanced countries, particularly the United States, Japan, and the member states of the European Union, where the public demand for safe products typically outruns the requirements of the international standards that govern trade in a specific product. Most often regulatory equivalence is established among members of a preferential trade agreement (PTA): a regional trade bloc whose members impose higher standards on themselves and exclude goods from nonmembers that do not meet them. The European Union is an exemplar in this regard, having expressly established agencies to use contrasting, national regulatory practices to establish and continually update unitary frameworks for equivalence in the single market, such as in food safety.

But the WTO, as the guardian of a single, integrated trading system, has made it difficult to form PTAs. WTO rules require that PTAs involve "substantially all" the trade among potential members. In effect, the agreements are meant to be as comprehensive regionally as the WTO aims to be globally. PTAs are agreed on after years of negotiations and countless horse trades, usually conducted in secret under expedited or "fast-track" conditions to make it difficult for damaged interests to coalesce against the deal. This secrecy, though, has made PTAs the emblem of the unaccountability of large trade deals. Negotiations that were unwieldy to begin with are frequently today a political impossibility.

All together, these developments—the obstacles that consensus decision-making poses for responding to nontariff measures, the intrinsic difficulty of entering into PTAs, and the political opposition they currently provoke—are redirecting efforts to reform the world trade order. With much of the WTO hamstrung by geopolitical conflict and the weight of outlived routines, the initiative is passing to small groups of willing innovators, concentrating on adjustments in sectors or problem areas rather than comprehensive agreements among many actors. At the WTO's December 2017 Ministerial Conference in Buenos Aires, this shift became public as groups of member states launched "joint statement initiatives" in e-commerce, the domestic regulation of services, investment facilitation, and measures to build the capacity of micro, small, and medium enterprises to utilize trade opportunities. The activists were small, sophisticated countries such as New Zealand, Switzerland, and Singapore, which live by trade and have often been at the forefront of movements for the reform of international standards.[9] Parties to these initiatives will restrict the imports of goods that do not meet the requirements they are setting without first obtaining the approval of all 164 WTO members, as the organization's rules would normally require.

The Buenos Aires meeting thus quietly marked the same break with top-down, consensus decision-making celebrated in climate change two years earlier in Paris. But where Paris, as we saw, ended by declaring the bankruptcy of the Kyoto approach, leaving the construction of an alternative for later, Buenos Aires went further; in practice, the joint statement initiatives are concretizing a new kind of trade instrument—the open plurilateral agreement (OPA)—that can address the accountability deficits of globalization and revitalize the WTO. This same platform could be used to construct sectoral, experimentalist regimes for decarbonization.

Open Plurilateral Agreements: Resolving the Trilemma

OPAs are bi- or multilateral agreements authorizing regulatory authorities in a particular domain to open the domestic market to the goods or services of other member countries—provided that the traded goods meet evolving national standards and that countries that do not yet meet the shared requirements may become parties to the agreement when they do.[10] Beyond breaking with the WTO's consensus requirement, OPAs differ from PTAs in three key ways, each of which can help restore public confidence in trade negotiations under WTO auspices.

First, OPAs are domain specific, not comprehensive trade deals like PTAs. There is no possibility for a lobby to protect its interest in one sector by securing a concession in another, and because such horse trades are impossible, there is no reason to guard against them by keeping trade negotiations secret. If public values are being sacrificed for commercial interests, the public will see it happening and can react.

Second, where PTAs are detailed agreements—fixing the terms of trade for the foreseeable future, subject to marginal adjustments—OPAs establish frameworks for continuing the reciprocal review of existing regulatory standards and their implementation. These institutional routines will bring the unexpected and troubling effects of trade agreements on communities and industries to light, or at least make it less likely that elites, in a repetition of the recent past, are more or less accidentally inattentive to the consequences of their advocacy.

The third difference between PTAs and OPAs concerns their openness to new members. To accede to a PTA, a new member must agree to all the terms of the elaborate compromise struck by the original signatories. PTAs are thus in practice closed agreements.[11]

By contrast, because they are domain specific, accession to OPAs requires of a candidate country only the narrower commitment to meet the regulatory requirements that apply to the particular class of goods and services for which it seeks market access. That commitment, too, can be tailored to reflect the candidate's circumstances. As members of an OPA require only equivalent performance—not identical procedures or institutions—in conformance testing, standards setting, and enforcement in each domain, candidate members can achieve the required regulatory outcome by the process best suited to their own traditions and conditions. Similarly, they can enter an OPA stepwise—establishing the equivalence of methods in one product, then another, and another—so that trade expands and collaboration deepens even when the full equivalence of regulatory systems is a distant goal. To aid the entry of new members, OPAs also provide technical support. Under the Trade Facilitation Agreement, a precursor to the joint statement initiatives concluded in 2013, developing and least developed countries falling behind the implementation deadlines for improving customs practices can evaluate their own difficulties. An expert secretariat then reviews this self-assessment to determine eligibility for support.[12] The procedures are similar to those used by the MLF to support the transfer of ozone-safe technology to developing countries under the Montreal Protocol and could, as noted in the last chapter, serve as a model for the provision of technical adjustment assistance under the Paris agreement as well as in OPAs.

Acknowledging the legitimacy of OPAs could gradually transform the nature of the WTO itself. Today the WTO is a forum for preparing the (re)negotiation of comprehensive trade deals that arouse public suspicion, with a dispute resolution system in the Appellate Body, whose legitimacy is questioned by key member states. This could change if regulatory differences continue to gain importance in trade agreements and the WTO becomes hospitable to OPAs—not just recognizing their legality, but reviewing and guaranteeing the integrity of their operation, and creating facilities for technical support services to candidate members. For example, the Appellate Body could review decisions by the current members of an OPA to reject a candidate's methods as not equivalent to their own; the precedents thus established could accumulate into a jurisprudence assuring that OPAs are truly open to new accessions. In this way, a WTO hospitable to OPAs could, gradually to be sure, become less the impresario of suspect deals administered by its own technocrats and more the guarantor of trade relations that nations enter into deliberately, knowing that misjudgments can be caught and corrected.

These three features also make OPAs ideal platforms for the development of experimentalist decarbonization. First, in both trade and climate regimes, the basic organizational unit is the sector—conceived of as a flexible problem-solving unit whose exact configuration depends on the (changing) nature of the problem to be solved. For some purposes and in some circumstances, for example, steel, cement, and fertilizer might be grouped in a single OPA. Second, the framework character of OPAs—and their congenital openness to revision—make them particularly well suited to the collaborative, technology-forcing standard setting at the heart of experimentalist innovation at the frontier. In both trade and climate regimes, standards support both coordination and exploration, and they are made to be revised. Finally, the support and new member provisions of OPAs resonate with the experimentalist destabilization of the status quo through penalty defaults in coordination with technical aid. Trade and climate regimes recognize that mastering novelty is hard. It must be learned rather than simply willed; there will be mistakes and delays along the way.

Beyond Sovereign Self-interest

Piecing together a trade regime from sectoral OPAs points then to the beginning of a resolution of the globalization trilemma asserting the incompatibility of national states, democracy, and orderly global trade. A regime of OPAs breaks with the uniform rules of hyperglobalization. There is no single, global regulator which, if technocratic, displaces democracy, or, if democratic, displaces the nation-state. In globalization by OPAs, regulation is negotiated market by market, under conditions that allow and encourage each state to make deliberate, democratic decisions about whether and under what conditions to trade.

But in restoring sovereign control over participation in global trade, a regime of OPAs does not fully return to a world of jostling sovereign nations, connected only by the temporary coincidence of their interests and constantly at risk of sliding into unregulated chaos. Invited to join an international order that it can in some measure shape to its condition and ambitions, a nation is called to consider what it could and will be as much as what its history and values command. The burdens and opportunities of interdependence can thus enlarge the traditional understanding of sovereign self-interest. The sovereign state is prodded to see itself neither as an unconstrained actor, answering at best to its people, nor as part of a postnational cosmopolitan order. Instead, it is part of a community of nations struggling to make sense of their differences as they define common projects.

The searching process by which China came to adhere to the Montreal Protocol in the early 1990s, described in chapter 2, was a flash-forward showing of how engagement with an accommodating world can favor this enlarged understanding of national self-interest. With the Chinese government itself deeply divided on joining the treaty, the national environmental regulator was asked to convene a comprehensive review. Finding that adjustment would be valuable but costly, and that phasing out ODS would strain the government's capacities, China made its agreement to the treaty conditional on the provision of financial support coupled with technical assistance in institution building. Those demands, backed up by similar demands in many other emerging countries, culminated in the creation of the MLF and its provision for collaboration in improving the "policy" and "institutional" frameworks of developing country parties to the treaty. What China, with its substantial resources and foresighted government, did exceptionally would become more universally accessible and routine in a world of OPAs, especially if support is provided in the early phases of the adhesion process.

We understand of course that the availability of more accommodating trade and climate regimes does not ensure their acceptance. No one today needs to be reminded of the ease with which democratically elected governments reject cooperation with outsiders as a betrayal of their values and a threat to their integrity. But even if some countries summarily reject OPAs as another ruse of technocracy, others will welcome, or be unable to avoid, the opportunity to ventilate domestic political differences; having gone that far, some will convince themselves to join the new arrangements. Their example will make it harder for subsequent holdouts to dismiss discussion out of hand. A regime that creates the pressure and opportunity for such national discussions is, in any case, much to be preferred to the current one, which all but incites rejection by forcing a choice between cooperation at the price of national values or the assertion of those values at the price of isolation.[13]

The trade regime and the climate change regime are at the same crossroad. Global solutions based on uniform rules under the direction of technical elites have failed in both. Our hope is for an alternative, pieced together from partial, more nearly local experiences, and more accountable and effective than the current regimes because it encourages deliberate, broad, and collaborative searches for solutions to urgent, concrete problems—and because it recognizes that those solutions too will need to be corrected in light of the experience of those who live under them. This book has tried to give a name to this organized combination of open-minded determination and humility in our dealings with the planet and one another: experimentalism.

NOTES

Chapter 1. Introduction: Toward Experimentalist Governance

1. Benedick 1998; Parson 2003.
2. Hoekman 2016; Hoekman and Sabel 2019.
3. Dewey 1927, 207.
4. Cullenward and Victor 2020.
5. Experimentalism shares with the work on governing the commons, associated with Elinor Ostrom, the assumption that local knowledge is indispensable to the solution of a broad range of complex collective action problems. Ostrom 1990.
6. In this sense, our arguments overlap with the literature on adaptive governance and active learning. Lee 1993; Social Learning Group 2001.
7. Alter and Meunier 2009; Drezner 2009; Raustiala and Victor 2004. Related are literatures that see the process of governance being forced open to accommodate many different groups. See, for example, Hoffmann 2011. That study also points to the value of experimentation, although it has a causal logic around experimentation quite different from ours.
8. Hale, Held, and Young 2017; Biermann et al. 2009; Abbott et al. 2012.
9. Barrett 1991. For a time, these debates played out as well through proposals for a World Environment Organization. See the review in Biermann and Bauer 2005.
10. Such surfeits exist in many areas of international governance and help explain the observation, made long ago, that international governance often takes the form of a "regime complex" of partially overlapping regulatory regimes. Alter and Meunier 2009; Drezner 2009; Raustiala and Victor 2004. Extending these ideas, including with application to climate, are many studies such as Hale, Held, and Young 2017; Biermann et al. 2009; Abbott et al. 2012; Keohane and Victor 2011.
11. In this sense, we resonate with Mazzucato 2013 in showing that experimentalism (in this study, a competent state steering the economy) is alive and well in what is widely thought to be a least likely case.
12. For a complementary view of dynamic politics, see Mallet and Khalaf 2020; Lauber and Mez 2004. For this logic applied, for example, to the cluster of "direct air capture" technologies, see Meckling and Biber 2021.
13. Oreskes and Conway 2011; Stokes 2020; Mildenberger 2020.

Chapter 2. Lessons from the Path Not Taken: Montreal and Kyoto

1. The agreement built on a "framework" agreement from two years earlier: the 1985 Vienna Convention for the Protection of the Ozone Layer. It was purely a framework agreement and contained no substantive commitments, and thus most experts see the Montreal Protocol as the

onset of real cooperation around the ozone layer. UNEP 1985. For fuller diplomatic histories, see Benedick 1998; Andersen and Sarma 2004; Parson 2003; Hoffmann 2005.

2. For the treaty, see UNFCCC 1992. For a fuller history of climate diplomacy, see especially Bodansky 1993.

3. For example, one study published at the completion of the UNFCCC specifically outlined the similarities with Montreal regarding the formal goals and procedures for adjusting treaty commitments in light of new scientific and technological information. Weiss 1993, 688.

4. See, for example, the comments of Winfried Lang. ESDO 2018. The definitive US-focused diplomatic history of the ozone talks, by Richard Benedick (the chief US negotiator on ozone during the formative years), identifies at least eleven major international environmental institutions, including the UNFCCC, that were heavily influenced by the ozone model. Benedick 1998, 331.

5. Victor 2011.

6. For an essay in counterfactual history on a large scale, see Sabel and Zeitlin 1985.

7. Perhaps the best articulation of the merits of a global market response was offered by someone who wasn't in Kyoto doing the diplomacy but nonetheless had the big picture of how governance was changing: Janet Yellen, at the time (early 1998) head of the Council of Economic Advisers. She testified in support of the recently drafted (late 1997) Kyoto Protocol. Yellen 1998.

8. Victor 2011.

9. Only four developing countries (Egypt, Kenya, Mexico, and Nigeria) actually joined the original Montreal Protocol. Three were special cases: Egypt was the home country of Mostafa Tolba (the United Nations Environment Programme's [UNEP] boss at the time and one of the key diplomatic entrepreneurs in the Montreal negotiations), Kenya hosted UNEP's headquarters and thus needed to show solidarity, and Mexico was the home country of Mario Molina, a favorite son given that he was one of the two scientists who discovered the impact of chlorofluorocarbons (CFCs) on the ozone layer. In crafting the original Montreal Protocol, the G77, the club of nearly all developing countries, was not active, nor were China and India. The treaty did recognize the need for a plan regarding these countries, defined the terms under which a country would be considered "developing" (Article 5), and deferred until later provisions on technical assistance to these countries (Article 10). Developing countries, so defined, were allowed a ten-year delay on any control measures. In effect, at Montreal the developing counties were directed "primarily to maintain maximum usage of CFCs for the longest possible grace period" and not much beyond that. Quoted in Benedick 1998, 148. But when the developed countries started to tighten controls, the developing countries needed a more elaborate plan.

10. Moreover, the UNFCCC was negotiated in the context of the 1992 Rio Earth Summit that focused heavily on the problems of sustainable development in emerging countries. Rio was organized, in no small part, around the increasing political prowess and diplomatic organization of developing countries, their demands that environmental protection benefit development, and their corresponding wariness of measures that would impose costs and constraints on their development pathways.

11. On the then-current understanding of ozone depletion, thinning—but not a hole—should have occurred, if at all, not in the cold of the Antarctic night but rather in the sunny middle latitudes, where solar energy could initiate ozone destroying in the upper atmosphere. Even when the science came to understand how reactions on the icy surfaces of clouds could destroy ozone at alarming rates at the dark wintery pole, calculations of the magnitude of human damage to the stratosphere diverged enormously. Models of ozone depletion only began tracking observed patterns of depletion in the mid-1990s, nearly a decade after the first treaties were agreed on. Parson 2003, 55.

12. Parson 2003, 55–56.

13. Benedick 1991, 83–88.

14. On "environmental conscience," see Ivanova 2021. On the central role of UNEP as the orchestrator of environmental cooperation at the time, see Tolba and Rummel-Bulska 1998. The huge diversity in experience at UNEP (and other international bodies working on environmental issues) yielded many alternative models for cooperation and was the subject of a major review of "lessons learned" in international environmental cooperation by Peter H. Sand, the principal legal officer for the 1992 Rio Conference on Environment and Development. Sand grapples with how to keep international cooperation from getting stuck in consensus, and encourage firms and governments to push frontiers. Sand 1991. The Sand analysis was cited by academics, but it was not heeded by diplomats as governments raced ahead to craft agreements on topics like climate change that superficially emulated the Montreal Protocol yet lacked understanding of how Montreal worked.

15. Bodansky 1993, 23; Haas, Levy, and Parson 1992.

16. Barrett 2005.

17. Najam, Huq, and Sokona 2003.

18. Andersen and Sarma 2004, 44.

19. US Congress 1977, section 126.

20. Andersen et al. 2018.

21. Andersen and Sarma 2004, 44.

22. Parson 2003, 53–55, 193. For the case for looking to the potential benefits of regulation, see Oye and Maxwell 1994.

23. Benedick 1998.

24. Under the 1977 Clean Air Act Amendments, Congress funded periodic scientific assessments, and the National Academy of Sciences played a central role. Conveniently for historians, the periodic academy studies—which became a partial blueprint for a systematic assessment of the ozone in the 1980s (which in turn became a partial model for the international assessment of climate change)—documented changes in the alarm level of the "consensus" over time. Among the many changes in ozone science were updates in the reaction rates for key steps in ozone-destroying catalysis, which in turn led to updates in ozone-thinning models. By the early 1980s, the trend was toward projections for lesser ozone depletion than originally feared. National Research Council 1983.

25. The talks were slow and meandering. Formally, the UNEP Governing Council authorized talks on a convention about the ozone layer in 1980 (Decision 8/7B); those talks began in 1981.

26. UNEP 1985; Lang 1991.

27. Sand 1991.

28. Parson 2003.

29. Sunlight still played a role, and that is why the hole appeared in the spring (October for Antarctica) just after sunrise. Cold was a key ingredient, for it allowed the formation of polar stratospheric clouds. On the role of polar clouds as the sites for the special heterogeneous chemistry that causes rapid ozone loss in the Antarctic, see Hofmann et al. 1989. For a key early paper on the depletion of the Antarctic ozone, see Solomon et al. 1986.

30. A huge thanks to Ted Parson, who has been generous with his insight as we learned how the Montreal method really works. What we call an "experimentalist" approach overlaps with what Parson calls an "adaptive regime." For a discussion relating adaptive thinking in environmentalism to experimentalism, see Cosens et al. 2020.

31. UNEP 1987, Article 6.

32. In addition, a distinct panel looked at the effects of ozone thinning on humans and the environment. But as the Montreal Protocol evolved, most of the stocktaking functions were done principally by the Scientific Assessment Panel, which was focused not only on the evolving state of the science but also the broad options for cutting ODS. The ozone hole revealed that even at low concentrations of chlorine and bromine in the upper atmosphere, there would be a lot of

harm to the ozone layer, and thus de facto the goal for the ozone regime became approximately zero human-caused chlorine and bromine in the upper atmosphere.

33. Formally, any changes that would alter the substantive commitments of the protocol—such as changing phaseout schedules or allowing time-limited exemptions—required a vote of the members of the protocol. The TEAP/TOC process provided the information needed to frame those votes and gained its influence, formally, by the fact that its recommendations nearly always had the largest impact in shaping the votes.

34. To underscore the fluidity of the committees, at first TOCs were established in refrigerants, foams, solvents, aerosols, and halons with a large number of members, yet little clarity around who would do what. Initially, in fact, they weren't TOCs at all but rather teams of authors charged with writing chapters in an assessment report. See UNEP 1989a. For more detail on balance across the five areas that became the TOCs, see UNEP 1989b, Annex V, 5. At the outset, TEAP wasn't TEAP but instead two panels with overlapping functions that later would be fused.

35. Parson 2003, 190.

36. Parson 2003, 189.

37. The US Air Force was a member of the Industry Cooperative for Ozone Layer Protection too since fully half of CFC solvents usage was mandated by US military specifications. The US military's participation meant changes for allies, and the US Department of Defense was soon coordinating technologies with militaries from Australia to Canada to the United Kingdom, and getting the North Atlantic Treaty Organization to endorse the shift. With the help of the EPA, the Department of Defense also cooperated with the Soviet military to elicit similar changes in their alliance. Andersen and Sarma 2004; Parson 2003, 190.

38. In 1988, just a few months after Montreal, the keynote speaker at the first conference was Montana senator Max Baucus, a central figure in US national regulation of ozone pollutants. While the topics under debate, such as refrigeration compressors and CFC-free soldering, were obscure, the deliberation occurred under the shadow of regulatory policy. Baucus 1988.

39. Parson 2003, 188–90.

40. Parson 2003, 190.

41. Christensen 2016.

42. Formally, the exemptions are granted by the parties to the treaty because they are time-limited alterations of the treaty obligations. But in every case, the party-decided exemption process begins with (and is framed by) the expert assessment described here. Generally the TOCs and system of panels have this framing function with regard to "adjustments" and "amendments" to the treaty. Adjustments are needed when the existing rules are tightened or loosened; amendments are necessary to change the scope of the treaty, such as by subjecting new chemicals to regulation. In both instances, the parties defer almost completely to the judgments of the panels in deciding whether to formally consent to the proposals.

43. Victor and Coben 2005.

44. Parson 2003, 194–95.

45. Parson 2003, 193.

46. As the science evolved, it became clear that healing the ozone layer would require essentially zero emissions of any chlorinated or brominated compound. The ozone hole was triggered at even low concentrations of chlorine and bromine in the upper atmosphere.

47. Benedick 1998, 163.

48. Zhao and Ortolano 2003, 710.

49. Biermann 1996.

50. Benedick 1998, 152–57.

51. On the funding mechanism, see especially Biermann 1996; DeSombre and Kauffman 1996. In China, one initial estimate put the cost of ODS controls at $1 billion while a preliminary

estimate for India imagined funding at double that level. Such numbers, anchored in no serious analysis, were far higher than donors were willing to pay. Benedick 1998, 187. The India estimate is reported in Parson 2003, 203, along with an estimate of $4 billion over a decade for all developing countries.

52. Greene 1998, 101–5.

53. Benedick 1998, 247.

54. Zhao and Ortolano 2003, 714, 717.

55. Zhao and Ortolano 2003, 717.

56. Benedick 1998, 247.

57. Andersen et al. 2018, 13–15.

58. Benedick 1998, 207.

59. Zhao and Ortolano 2003, 713.

60. Benedick 1998, 207–9, 220–30.

61. Parson 2003, 211, 218, 227–28. Among other things, the first full assessment of methyl bromide control options finished in 1994 revealed why the first TOCs had been so effective: they were generally smaller, more focused, and in the case of CFCs, didn't include producers. The Methyl Bromide TOC, with a membership of sixty-five, included manufacturers and some users who steadfastly "fought to have the report conclude that there were no alternatives to [methyl bromide]." Quoted in Parson 2003, 228.

62. MBAO, n.d.

63. UNIDO 2015.

64. As climate diplomacy was gathering steam from 1989 to 1991, the Austrian diplomat Winfried Lang—known as the father of the Vienna Convention and the man who helped organize early European support for cooperation on the ozone layer—singled out the ozone depletion regime as the model to be emulated for the climate, though he allowed that progress on climate change would be slower. His account of the workings of the ozone regime did not mention the TEAP or TOCs; industry, where it figures at all, is noticed for its opposition to the regime rather than its engagement in the joint search for solutions. Lang accurately reflected what most architects of environmental cooperation thought at the time. Lang 1991.

65. A historian of climate diplomacy, Dan Bodansky, dates the end of Western nations' dominance on the climate agenda to about 1990. Prior to that debate, much of global diplomacy was focused on control strategies and debates among Western countries. After that, developing countries were much more organized and influential. See Bodansky 2001, 28, 30.

66. Hecht and Tirpak 1995. For the first Intergovernmental Panel on Climate Change report, see IPCC 1990. Still other countries saw the World Meteorological Organization, the host of the 1990 Second World Climate Conference, as a logical host for climate diplomacy. Bodansky 2001.

67. In addition to the traditional voice of developing country interests—the G77, which typically caucused and spoke in tandem with China—a new subset of these nations formed in 1990: the Alliance of Small Island States. The group, representing the countries most vulnerable to a changing climate, was so effective that it soon became recognized as an official grouping of nations within the UN system. Ronneberg 2016.

68. UNGA 1988, 1990.

69. Vogler 2007. On the relation of the newly independent developing countries to the former colonial powers and their place in the global economy, see Getachew 2019.

70. Bodansky 2001.

71. Bodansky 1993.

72. UNFCCC 1995.

73. The main decision from Berlin was called the Berlin Mandate, and it set out the process and expectations for those negotiations. UNFCCC 1995, 1/CP.1. Formally, targets and timetables

were called, in the language of the Berlin Mandate, "quantified emission limitation and reduction objectives" (QELROs)—a reminder that even the simplest idea can be conveyed by mangled, complex language.

74. UNFCCC 1995, 1/CP.1. This exemption was celebrated by many because the Berlin Mandate thus gave precise, institutional meaning to the principle of "common but differentiated responsibilities and respective capabilities" introduced in the UNFCCC four years earlier. Rajamani 2013. The sharp distinction between country groupings was taken to the limit in Kyoto, which rejected proposals that would have allowed developing countries to make even *voluntary* commitments to emissions reduction.

75. For the Montreal language, see the operative provisions of Article 6 in UNEP 1987. For the UNFCCC, see the authorizing language for the Subsidiary Body for Scientific and Technological Advice in Article 9 in UNFCCC 1992. In chapter 6, we will show that even where technically oriented meetings have been held, the focus has been diplomacy, not technical analysis.

76. Indeed, these same groups had evaluated and rejected concepts like "pledge and review" that might have evolved into a more experimentalist system of governance for climate change. Targets and timetables—without opportunities for revisions and the coadjustments of pledges in light of implementation experiences, were the centerpieces of their demands at Berlin. Victor 2001.

77. For the origins of the enthusiasm for emissions trading starting in this period, see Hahn and Hester 1989; Boyd 2021. For a discussion of the limits of the market-based approach to problems of this type, see Sabel and Simon 2011, 53. For a discussion of market mechanisms when applied to the rigidity of US environmental law, see Ackerman and Stewart 1984.

78. On the real-world experience with bubbles and other early market-like concepts, see Hahn and Hester 1989. See also Stavins 1988. On the sulphur program, see chapter 4.

79. Boyd 2021, 460. Advocates of emissions trading were well aware of the hot spot problem and suggested that it could eventually be addressed through elaborate refinement of the basic scheme. Sabel and Simon 2011.

80. Stewart and Wiener 2004.

81. Yellen 1998.

82. Victor 2001.

83. Victor 2001; Cullenward and Victor 2020, 87–102.

84. Streck and Lin 2010.

85. While the problem of CDM quality attracted press attention in the United States as an example of a UN program gone awry, in Europe concern was even greater because CDM credits were flooding into the ETS and were one factor in eroding prices in the ETS—in effect, Gresham's law. For work emblematic of the European Union's investigation into CDM quality, see Cames et al. 2016. That study, while it appeared after the key decisions were made, is a good overview of the methods and issues that EU policy makers were grappling with, and that led to the conclusion that the CDM can't be considered a single mechanism; rather, each project type and institutional setting required distinct analysis. Staff work summarized around 2010 outlined a series of reforms to the CDM itself that might have improved quality (we are skeptical because the problem of setting baselines is intractable, no matter how clever the reforms), but such reforms would have required changing the CDM itself (which would require diplomatic consensus) or erecting a parallel EU system to oversee the CDM (which would have been administratively impractical). European Commission 2010. For the EU policy that has curtailed usage of the CDM in Europe, see EU 2013.

86. Victor 2001.

87. Canada and Australia were in similar situations. Canada joined Kyoto, but then withdrew when its largest trading partner (the United States) did not join. Australian politics on climate change were similar to, if not more toxic than, those in the United States, which made joining Kyoto impractical.

88. Victor and Salt 1995.

89. Revkin 2009.

90. In China that decade, the carbon intensity of the energy system actually went up—coal grew disproportionately faster than all other energy sources—and emissions soared. Victor et al. 2014.

Chapter 3. Theory of Experimentalist Governance

1. Chandler 1993.

2. Christensen 2016.

3. On the distinction between mass production by specialized tools and flexible production using reconfigurable capital goods under the control of skilled workers, see Piore and Sabel 1984.

4. From this perspective, decarbonization is a key piece of the larger program of overcoming the dualist separation of the economy into an advanced sector—the knowledge economy—that thrives on innovation—and a stagnant, rearguard sector that creates dead-end jobs and blighted communities. The Green New Deal aims to connect the two projects, but details are scarce. For a proposal to create more good jobs, extending the scope of the advanced sector, in line with the thinking here, see Rodrik and Sabel, forthcoming.

5. But this illustration suggests that the incentive-design problem is easy to solve. In fact, it turns out to be formally unsolvable. The principal's best strategy is to in effect bribe agents into disclosing one another's true costs of alternative actions by offering them shares in the project's returns. The only scheme that generates incentives large enough to induce disclosure, however, also leaves the principal with the discretion to cheat agents by withholding the promised payoffs, undercutting cooperation before it begins. See Holmstrom and Milgrom 1991; Holmstrom 1982. Actors in the real world solving principal-agent problems wing it.

6. Fama and Jensen 1983.

7. On stakeholder views, see Henderson 2020. On the failures of current corporate governance and the possible advantages of the private equity model, see Gilson and Gordon 2020.

8. The distinction between risk and uncertainty was developed by the economist Frank Knight in the 1920s. See Knight 1921.

9. Since the 1990s, a wave of "new" governance writing has called attention to the spread of loosely defined, posthierarchical structures, often in the form of networks, and frequently without explicit reference to the breakdown of principal-agent relations. See, for example, Rhodes 1997; de Búrca and Scott 2006; Scharpf 1997. Much of this literature focused on developments in the European Union, where the push to create the single market led to the creation of regulatory agencies that had to pool the experience of the member states and continue to learn from their differences while maintaining common frames for action. Experimentalist governance developed as a response to changes in the United States and was then applied to an analysis of innovations in the European Union. See Dorf and Sabel 1998; Sabel and Zeitlin 2008.

10. On the logic of mass production and its vulnerability to shocks, see Piore and Sabel 1984.

11. Smith 1976, 21–25.

12. Ben-Shahar and White 2006.

13. Trebilcock 2017.

14. Ward et al. 1995.

15. Womack, Jones, and Roos 2007.

16. Ausubel, Wernick, and Waggoner 2013.

17. For example, hazards to supply chains. See Kinghorn 2016.

18. For a prominent case study of the reaction of the US nuclear power–generating industry to the Three Mile Island disaster, see Rees 1996. On incident reporting in civil aviation, see Mills and Reiss 2014.

19. Holling 1978; Lee 1993.

20. Gilson, Sabel, and Scott 2009, 2014.

21. This section draws on Kessler and Sabel 2021.

22. The legal terminology is confusing. The Administrative Procedure Act of 1946 calls the traditional method of rule making by hearing, with presentation of evidence as in a courtroom, formal, and notice-and-comment rule making informal. In practice most rule making is of the notice-and-comment type and it has become the de facto standard of formality. In what follows we mean notice-and-comment rule making when we speak of formal procedures in contrast to the informality of guidance. See 5 U.S.C. §553.

23. Elliott 1992.

24. Vermeule 2016.

25. Parrillo 2019, 168–169.

26. See, for example, American Bar Association 1993. Nicholas Parrillo speaks in this connection of "principled flexibility." See Parrillo 2019.

27. Parrillo 2017, 34.

28. Parillo 2017, 34.

29. Parrillo 2019, 167–68.

30. Lindsay 2018.

31. Kessler and Sabel 2021 suggests some doctrinal and institutional reforms in this direction.

32. Callon, Lascoumes, and Barthe 2009.

33. For a concise portrayal of this skeptical literature, see Fishkin 2018.

34. Sunstein and Hastie 2015.

35. For a concise description of how deliberative polling is organized and the role of experts in providing sound information to frame discussion, see Luskin, Fishkin, and Jowell 2002, especially 458–60.

36. On deliberative polling in the radically polarized circumstances of Northern Ireland, just after it emerged from outright violence between Protestants and Catholics, see Luskin et al. 2014. There is, as can only be expected, disagreement among those studying deliberation as to whether deliberative polling reliably proxies what public opinion would be under plausibly favorable conditions. See Neblo, Esterling, and Lazer 2019.

37. See, in general, Fishkin et al. 2021.

38. Ho 2017.

39. Ho 2017, 43.

40. To eliminate the possibility that background features of the situation—the participants' length of service perhaps, or the experiences of one or another group—might in operationalizing the experiment come accidentally to shape the results as well as to enable the precise measurement of the effects of reform, the entire staff was randomly assigned to a treatment group that would conduct the weekly peer reviews and a control group that would continue inspection as usual. To further reduce the likelihood of accidental, extraneous influence, the peer pairings were randomly reassigned each week, with restrictions to ensure that pairs did not repeat and that assigned establishments were outside the pairs' home inspection territories. Within each pair, the peers alternated in the role of lead inspector, who took primary responsibility for guiding the visit, and whose report, corrected in light of discussions after the inspection, became the official record of violations, while the "nonlead" inspector filed an unsigned report coded for the purposes of the experiment as peer review. A great advantage of this procedure was to allow Ho to trace the evolution of judgment inspector by inspector and week by week. Ho 2017, 31.

41. Ho 2017, 45, 54.

42. Personal communication, Daniel E. Ho, December 8, 2021.

43. Ho 2017, 93.

44. Ho 2017, 57.

45. Goodin and Niemeyer 2003 emphasize the role of exposure to new information, credibly presented, in promoting reconsideration of views.

46. Ho 2017, 92, 86, 67.

47. Ayres and Gertner 1989.

48. Hiscox and Smyth 2006.

49. Gunningham, Kagan, and Thornton 2004.

50. Copeland 2012.

51. Clean Air Act 1963.

52. On the California effect, see Vogel 1997. On the Brussels effect, see Bradford 2020.

Chapter 4. Innovation at the Technological Frontier: Three Policy Icons and a Common Approach to Uncertainty

1. Gardner 2014.

2. Under Section 177 of the Clean Air Act, states can opt to follow California's clean air vehicle standards under a waiver granted by the EPA. For more, see Smith 2019.

3. California Health and Safety Code 1975, *§39500, §39510*. See Reed 1997.

4. CARB 1994. California has consistently failed to meet either standard. Bedsworth and Taylor 2007.

5. Hughes 2015.

6. Vogel 2018.

7. Reed 1997, 785.

8. International Harvester Co. v. Ruckelshaus 1973; Natural Resources Defense Council, Inc. v. US Environmental Protection Agency 1981.

9. Natural Resources Defense Council, Inc. v. US Environmental Protection Agency 1981.

10. Dixon, Garber, and Vaiana 1996, 7.

11. As one senior regulator put it in an interview, "[A manufacturer might say,] 'Look, I have this product, and I think it can be really cost-effective and I think it can really lower emissions' or 'I have the data to show you I can do it, and I can do it better than my competitors and I can do it cheaper.'" Wong 2015.

12. Cherry 2016.

13. Because of this expertise, CARB can "meet with manufacturers and we can roll up our sleeves and talk technical talk with them. It got us a lot of respect with them." McCarthy 2015.

14. CARB 1996, 4–5; Hughes 2015. See also Magnusson and Berggren 2011, 7–8: "Hence, systems integration still constitutes a critical activity in automotive development and design. Even though automotive manufacturers do not have to manufacture the complete vehicle in-house, they still need a broad technological knowledge base to be able to absorb new technologies (Cohen and Levinthal 1990), as well as to understand interdependencies in the product system and cope with imbalances due to different rates of development in different technologies (Brusoni and Prencipe 2001)."

15. One regulator described the push-pull dynamic created by CARB's feasibility requirement this way: "We have to know when we adopt the more stringent standards, there's a way to get there. And of course, they have to know there's going to be a market for the product they come up with." McCarthy 2015.

16. CARB 1996, 4–5.

17. Reed 1997, 722–23.

18. Reed 1997, 723.

19. Reed 1997, 733–34, 737–40.

20. BTAP 2000.

21. Reed 1997, 57.

22. Bedsworth and Taylor 2007, 7.

23. Reed 1997, 771.

24. Reed 1997, 774.

25. That year, all battery and plug-in hybrids accounted for 8.9 percent of new car sales; pandemic turmoil make data from 2020 less indicative.

26. Reed 1997, 774–75.

27. Bedsworth and Taylor 2007, 7–8.

28. Reed 1997.

29. Reed 1997.

30. *Rocky Mountain Farmers Union et al. v. Corey* 2013.

31. Dixon, Garber, and Vaiana 1996, 11.

32. CARB 2012.

33. For California sales, where about 8 percent of new vehicles are electric (including plug-in hybrid vehicles), see Kane 2020. And for global patterns, which remain robust despite big reductions in subsidies in China, thanks to expanded policy support in other parts of the world, see Holman 2020.

34. Chestnut and Mills 2005.

35. Environmental Defense Fund 2018.

36. "The Invisible Green Hand" 2002.

37. Ruckelshaus at the EPA's National Compliance and Enforcement Conference, January 1984.

38. Turner and Isenberg 2018.

39. Depending on the plant design, there will be other pollution devices as well—baghouses, filters, and in the most modern plants, chemical systems for removing nitrogen pollution.

40. Gypsum in theory has value as the key ingredient in wallboard, but only if there aren't heavy metals and radioactive particles in the gypsum. Most of the sulfurous slurry from scrubbers is simply a waste product—that is, another cost.

41. Barker 2012.

42. EPA 1990.

43. Environmentalists favored some data and ideally wanted to cut emissions nearly in half; industry did not mind a study program that delayed action; neither was powerful enough to force its preferred solution on the other. Formally, the US approach to acid rain looked like a rational study-then-regulate program. In practice, the arrows of causation ran in all directions, and the science played almost no role in shaping the politics. The science that could have mattered—estimates of the cost for cutting sulfur—turned out to be wildly wrong because analysts did not update their models to reflect a trove of rapidly shifting factors such as improvements in scrubbers, expansion of coal leasing in the West (mainly Wyoming), and the deregulation of the railroads. All told, the United States spent $500 million on the National Acid Precipitation Assessment Program, which issued its formal report after the 1990 Clean Air Act was completed. The program sponsored a lot of good research on air pollution sources, transportation, and ecological effects. None of it affected policy because political pressures ran faster than the scientists.

44. Association of American Railroads 2011; Hecker 2007, 6.

45. Stavins 1988.

46. Newell and Rogers 2003.

47. Hahn and Hester 1989.

48. Schmalensee and Stavins 2017; Ellerman 2010.

49. Bedsworth and Taylor 2007.

50. Stavins 1988.

51. Turner and Isenberg 2018.

52. By the time the program was designed, the emissions from the covered plants were about sixteen million tons. Some plants, due to other pollution controls, were not covered. And the total US emissions of SO_2 was higher than twenty million tons due to other industrial sources, including sulfur in petroleum fuels used for road transport.

53. A draft report was circulating while the Clean Air Act amendments were finalized, but that draft offered no crisp advice on where to set pollution limits. An earlier "interim report" had been released in 1987, but it was pilloried as hewing to the Reagan administration's probusiness position and thus ignored politically.

54. Schmalensee et al. 1998.

55. Turner and Isenberg 2018.

56. Schmalensee and Stavins 2017.

57. The availability of those options varied with exogenous conditions as well—leading, as market theorists rightly expected, to variations in prices. In 2005, sulfur prices peaked at more than $1,200 per ton in response to a series of exogenous shocks, such as derailments of coal cars that made it harder to deliver low-sulfur coal.

58. Studies of patenting data show that during the era of command-and-control regulation, there was a much higher level of innovation in sulfur control than during the market period. Taylor, Rubin, and Hounshell 2003.

59. "Smog" is the word frequently used for photochemical ozone pollution in cities. But the same chemical processes that make ozone (which is harmful to humans and plants when contacted in the lower atmosphere, but good for ecosystems up in the stratosphere) also run regionally; source pollutants such as hydrocarbons and carbon monoxide mix with NOx and sunlight in the atmosphere, and make smog downwind and regionally. If a locality must meet smog rules (per the NAAQS) locally, it often can't have a full impact without getting cooperation from states upwind.

60. Schmalensee and Stavins 2013.

61. Chan et al. 2018.

62. Schelling 1983; Sabel, Fung, and Karkkainen 1999.

63. EPRI 2005; American Electric Power 2017.

64. Cohen and Noll 1991; Lester and Hart 2012.

65. For an exception that proves the rule, see Albert Hirschman's (skeptically received) suggestion that large development projects are governed by the principles of the "hiding hand": they are undertaken because their advocates underestimate the challenges to be faced, and succeed because difficulties, providentially, lead to the discovery of unexpected problem-solving capabilities. Hirschman 2014.

66. International Energy Agency 2012, 2020a, 2020b.

67. Chan et al. 2018.

68. National Academies of Science, Engineering, and Medicine 2017.

69. Bonvillian, Van Atta, and Windham 2019.

70. There is attention, certainly, to what potential customers might want; indeed, the early stages in creating a new DARPA program involve extensive canvasing of possible use cases. Bonvillian, Van Atta, and Windham 2019.

71. Rodrik and Sabel, forthcoming.

72. Azoulay et al. 2019.

73. Rodrik and Sabel, forthcoming.

74. Rodrik and Sabel, forthcoming.

75. Azoulay et al. 2019.

76. Howell 2017.

77. Doblinger, Surana, and Anadon 2019.

78. National Academies of Sciences, Engineering, and Medicine 2021b.

Chapter 5. Experimentalism in Context: Ground-Level Innovation in Agriculture, Forestry, and Electric Power

1. The idea that the final, decisive interpretation of price signals and other unambiguous guides to action is rooted in mute, local circumstance is itself deeply rooted in economics. The kind of statistics expressed in prices, economist Friedrich Hayek writes, "have to be arrived at precisely by abstracting from minor differences between the things, by lumping together, as resources of one kind, items which differ as regards location, quality, and other particulars, in a way which may be very significant for the specific decision. If we can agree that the economic problem of society is mainly one of rapid adaptation to changes in the particular circumstances of time and place, it would seem to follow that the ultimate decisions must be left to the people who are familiar with these circumstances, who know directly of the relevant changes and of the resources immediately available to meet them." Hayek 1949, 83–84.

2. Hirschman 2014. For an exception in the study of technological borrowing, see Yi 2020. For a view of implementation as reinvention in use akin to the one here, see Eric von Hippel's work on the innovative modification of software or capital goods, such as machine tools, through lead users in collaboration with one another or producers. von Hippel 2006.

3. Stokes 2020.

4. Hutchins 2000; Latour 1999.

5. The consolidation of Irish milk processing is less pronounced than in Ireland's major dairy export competitors, such as Denmark, the Netherlands, and New Zealand, where one company processes as much as 70 to 80 percent of the milk pool. Dairy Industry Prospectus Report 2003.

6. Eurostat 2017; Food and Agriculture Organization 2018, 5; Bord Bia 2016.

7. Fitzgerald 2019.

8. Thorne et al. 2017.

9. Alothman et al. 2019.

10. It is estimated that in 2008, the wider agri-food (which includes beverages, infant formula, and food ingredients) accounted for 40 percent of the total net foreign earnings of all primary and manufacturing industries. Riordan 2012.

11. The Food Harvest 2020 strategy of 2010 set a target by 2020 of a 50 percent increase in the volume of milk production over the average of 2007–9 milk supply (4.93 billion liters). The volume of milk production in 2017 had reached 7.27 billion liters—an increase of 47 percent. Central Statistics Office 2018.

12. Poikane et al. 2014.

13. For the guidance documents, see "WFD Guidance Documents," n.d. For a detailed discussion of the Common Implementation Strategy as an experimentalist institution at the heart of the Water Framework Directive, see Scott and Holder 2006.

14. Hering et al. 2010; Voulvoulis, Arpon, and Giakoumis 2017; Giakoumis and Voulvoulis 2018.

15. Shortle and Jordan 2017.

16. Burgess 2018.

17. Daly, Archbold, and Deakin 2016, 161.

18. For a good overview of the role of LAWPRO's Catchment Assessment Team and its place in the new governance structure, see EPA Catchments Unit 2018.

19. On the Agricultural Sustainability Support and Advice Programme, see "ASSAP—Farming for Water Quality," n.d.

20. For the most current review of LAWPRO in operation, commissioned by the Irish EPA, see Boyle et al. 2021.

21. Clarke et al. 2014. Exactly how far electrification must go is unknown. If other energy carriers, like hydrogen, prove cost-effective and reliable, then they may play a big role in deep

decarbonization. If not, then more of the work of deep decarbonization will need to be done by electrifying end uses and then decarbonizing the power sector. Azevedo et al. 2020a.

22. Among the studies looking at the early days of renewables, including in California, Gregory Nemet's is notable. Nemet 2019.

23. "SB-1078 Renewable Energy" 2002; Jacobson 2020.

24. Pyper 2019.

25. Jacobsson and Lauber 2006; Hockenos 2015.

26. Sivaram 2018.

27. Nemet 2019.

28. Jacobson 2020.

29. The governor's own Office of Planning and Research orchestrated the effort because no state agency had the authority to convene all the right actors, which included not just solar developers but also city officials, organized labor, and fire departments. (Solar panels, badly installed, were a fire hazard.) Governor's Office of Planning and Research 2014.

30. Governor's Office of Planning and Research 2012.

31. For example, different kinds of rooftop projects—different sizes and local grid configurations—required different kinds of interconnection and top-level rules to govern that process. It was feared, with some reason, that the local utilities that controlled the local grids were using risk-averse grid interconnection rules to prevent the implementation of rooftop solar. Changing and streamlining those rules was thought to advance implementation in the traditional sense. Governor's Office of Planning and Research 2014.

32. Annual Levelized Cost of Energy Analysis (LCOE 14.0) 2020.

33. We describe the macro patterns here, but the real world of course is more complex. In particular, there are scenarios in which large numbers of small rooftop generators have synergistic impacts that could propagate across the grid—for example, if a frequency disturbance propagates into rooftop systems and the "droop" set points on the inverters cause them all to trip simultaneously. (There are interesting scenarios that start with cyberattacks that lead to similar outcomes.) Such concerns, in part, have led some jurisdictions to make costly upgrades to the existing inverters and inspired a lot of attention by engineers to inverter stability at all scales. All that said, scale is more important—and what grid-connected solar offers is scale.

34. CPUC 2015.

35. Simulation occurs over multiple timescales. Some models are used for long-term planning; others run real time alongside the grid to help operators manage problems as they arise.

36. Roberts 2018.

37. CPUC 2015.

38. The full story is more complex, of course. There is a third agency, the California Energy Commission, which oversees policy strategy and works jointly with the other agencies to redirect investment for the long term; the commission also administers various funding mechanisms (as do CPUC and CAISO) to support advanced technologies.

39. Indeed, curtailment was assumed to be such a nonproblem that CAISO didn't track or report it systematically until 2017. Nonetheless for grid operators and planners accustomed to traditional thinking—that grid operators would simply find ways to integrate renewables, and any excess would be small and easily dumped—the curtailment levels of 2017 would seem acceptably small. Only as real-world deployment advanced in the following few years did the numbers climb steeply. Moreover, once modelers caught up with the speed of renewables deployment, they showed that a massive reliance on renewables alone would lead to extensive curtailment along with other side effects, such as huge land areas needed for the overbuilding of renewables and huge needless costs. On the curtailment data, see "Managing Oversupply" 2021. And for a recent look back at the problem of curtailment and overbuilding, see Cohen et al. 2021.

40. CPUC 2008.

41. Formally, the planning process is done by opening a proceeding through which utilities make proposals. For regulated utilities (known formally as "electric corporations"), CPUC can mandate participation in a proceeding and thus de facto procurement. For public power agencies, CPUC's authority is more limited, and therefore the law requires that those organizations go through their own local planning process and feed that information to a different state body. Given this, as a practical matter, the experimentalist approach to battery deployment has focused principally on the utilities because they interact most extensively with CPUC. "AB-2514 Energy Storage Systems" 2010.

42. CPUC 2013, 6.

43. CPUC 2013, 15, table 2. The question of battery deployments on the high-voltage system was outside CPUC's oversight—a matter for CAISO, which more slowly developed its own program.

44. CPUC saw its mission as "transformational" in creating a new market for batteries, not just deploying a quota of storage devices. Thus the original plan was to evaluate projects against the potential for transformative impacts, not simply near-term costs and benefits. See CPUC 2013, 6.

45. Letting such projects compete, CPUC said, "would inhibit the fulfillment of market transformation" that CPUC saw as a central purpose of its storage policy. See CPUC 2013, 34.

46. Personal communication, former executive at a major California utility.

47. On the use of experience to drive changes in procurement, see CPUC 2013, 25.

48. See CPUC 2013, 26. Moreover, as a recognition that local details mattered, CPUC adjusted target compliance utility by utility to reflect the pipeline of storage projects that utilities already had funded with support from the Self-Generation Incentive Program and other incentive programs. Thus top-level goals, even at the outset, were highly contextualized to local circumstances. Once CPUC began this process, it learned that lots of special projects claimed exemptions; hence, like the Montreal Protocol TOCs, it identifies top-level rules that would align with reason and then sets those rules, provisionally, as the metrics for compliance. CPUC 2013, 32. See also CPUC 2013, 27, section 4.5.

49. It was known as the "power charge indifference adjustment," which CPUC adjusted in a time-limited fashion (explicitly rejecting arguments that the changes should extend beyond a decade since CPUC had no idea how the battery market would evolve). See CPUC 2014. For a decision that looks back at the procurement experience and also addresses issues with value stacking, discussed below, see CPUC 2018, 3.

50. CPUC 2016.

51. The subsidy scheme is known as the Self-Generation Incentive Program and it paid some of the costs for advanced battery projects; CPUC adjusted, utility by utility, the procurement targets to reflect its assessment of which Self-Generation Incentive Program projects should "count" toward the goals—a detailed regulatory oversight that is reminiscent of what the Montreal TEAP does when it recommends time-limited exemptions for critical uses of ODS, discussed in chapter 2. CPUC also adjusted the rules to make it easier for Community Choice Aggregators to participate in battery deployments. On the former, see CPUC 2013. On the latter, see CPUC 2016. It is worth noting that CPUC clarified that Community Choice Aggregators must assure "resource adequacy" for their customers—that is, to have contracted enough power to supply expected loads (and then some). Resource adequacy is one of the major mechanisms through which regulators help assure network reliability, and the scramble to meet resource adequacy rules has led some Community Choice Aggregators and other power providers to invest in more battery projects (and projects that integrate batteries with solar generators) for resource adequacy purposes.

52. Fitzgerald et al. 2015, 7.

53. California regulators, based on the deployment experience in the state, convened a workshop in 2018 to evaluate what was being observed and learned. See CPUC 2018, appendix B.

54. In practice, there are several different types of interconnect agreements, depending on the voltage, configuration, and application of the device in question. One of the many concerns of regulators that are trying to create opportunities for new technologies is that incumbent utilities will use the interconnect process as a barrier to entry—thus the process is overseen strictly (as is the modeling that feeds into the interconnect agreement) with an eye to evenhandedness. Moreover, when a utility installs a battery on its own network—thus raising such concerns about self-dealing—there is a requirement for an interconnect between the arm of the utility that deploys the device and the arm that manages the grid—a requirement designed in part to reduce the self-dealing risks. Nonetheless, griping is rampant, especially when an interconnect process leads the utility to find that a costly grid upgrade will be required. One advantage of storage projects is that they have, in most cases, been designed to use the services of the battery system to avoid the need for grid upgrades.

55. Formally in California, these rate and cost recovery decisions are taken after the expenditure is made, and thus the procurer is especially attuned to the likely interpretation of rules and experience regarding the prudence of its spending.

56. The process we outline here is most extensive for the largest projects—those that typically offer the greatest value (and potential risks) to the grid. For the smallest projects, such as most behind-the-meter deployments, the utility and regulators have much less oversight, which helps explain why the behind-the-meter market has seen the most expansive growth.

57. It is also emblematic of the attention to how models (to a limit) can help advance the "adaptive management" of renewable resources. See Holling 1978.

58. The answer required navigating federal regulatory standards (which defined grid-connected resources) along with challenges in the power flow modeling of resources that were unlike anything that had been connected to the grid. CAISO first grappled with these issues in 2013 when it looked at the initial cluster of battery projects (thanks to CPUC's procurement mandates); at the time, there was essentially zero experience anywhere in the world to offer an empirical guide for the modeling, and hence the modeling techniques were developed from logic and comparing batteries with other, more familiar resources. See CAISO 2014.

59. For more on how these "non-generator resource" deployments might affect the energy markets, see CAISO 2011. For a discussion of how the interested parties affected the setting of interconnect rules—and how uncertainties in battery operation would affect the reliability of the grid, see CAISO 2014. And for more on the factors affecting deployment of batteries over time, see Hart and Sarkissian 2016.

60. That project was a large Tesla battery system located at a wind farm and put into service in 2018. For a study in 2017 that made predictions and focused on the battery adding value by arbitraging power, see MacKinnon 2017. For a retrospective on how the battery actually delivered value, which was with voltage support and other "ancillary services," see Parkinson 2020.

61. Indeed, worries emerged within the regulator that its battery program was not pushing enough experimentation. As the idea of value stacking was put into practice, it became easier to identify projects using the known values that were the most cost-effective. This was leading to an overinvestment in one class of technologies (notably lithium chemistry) and known applications around the medium-voltage grid. According to interviews, in the late 2010s, CPUC became concerned that this approach was narrowing the scope of search and the rules might need adjustment. Part of the problem was that the review process at CPUC was mainly backward-looking. The investor-owned utilities would advance their battery projects, and get approvals for rate base

and other costs only after committing the resources, and CPUC reviews of the whole program looked back after that. How to rectify this situation, which is distinct from the experience with most of the Montreal Protocol TOCs that also did forward-looking evaluations of projects, is an ongoing challenge.

62. CPUC 2018, 15–18.

63. The CPUC order that created rule 6 (and other rules) explicitly called these provisional, and established a process to investigate performance and make future adjustments, if needed.

64. See CAISO, CPUC, and California Energy Commission 2021. While the main troubles in August were not related to batteries, a large part of the analysis focused on how batteries providing resource adequacy, the key measure of whether a power supply is reliably available, actually functioned on the grid—a novel topic because the battery market was changing so quickly. (The resource adequacy provisions are one way that regulators make sure that at every hour of the day, there are sufficient generating resources on hand. In practice, the shift to renewables and the need for more power that can be dispatched quickly are forcing a rethinking of the whole concept of resource adequacy, and also forcing the regulator to develop new methods for measuring resource adequacy—methods that overlap, in concept, with the concept of value stacking; one of the many values to be stacked is power flow at specific moments in time, which is what resource adequacy measures.) Prior to 2018, CAISO and CPUC rules in effect did not allow batteries to provide resource adequacy. Until the operational experience under the first wave of procurement, there was neither the installed base nor the experience to craft effective resource adequacy rules for batteries. In August 2020, CAISO could observe, for the first time under conditions of severe grid stress, how the initial two hundred megawatts of batteries procured for resource adequacy actually performed.

65. See CAISO, CPUC, and California Energy Commission 2021. This study finds that batteries might have been able to provide greater reliability services than they actually were called on to supply in August. The conservative approach embodied in rule 6 may help explain that outcome. For more, see Penn 2020. Nonetheless, CAISO remains cautious—as any regulator with the responsibility for grid reliability must—because of the ways that the operation behavior of battery systems remains murky yet increasingly important as they get larger and more material to grid management.

66. One big open question is whether markets will be established for all or most stackable services, or whether investment will instead proceed principally under a rate-of-return regime. But regardless of how that question is answered, contextualization under a regulatory regime of the kind CPUC has devised is likely to continue into the indefinite future. The decentralization of power generation and management will continue to cascade within many of the (larger) local nodes of the grid, such as big-box stores, hotels, hospitals, or universities.

67. For an example of two projects underway, see Center for Sustainable Energy 2020. For a more systematic look, see the "ARD, S8" projects, summarized in California Energy Commission and Doughman 2020.

68. On the projects underway, see, for example, Center for Sustainable Energy 2020. Meanwhile, CPUC procurement has focused on a new frontier: encouraging the mass deployment of local networks of batteries and generators known as microgrids. The logic behind this push, which is destabilizing the status quo again, is that more local control and reliability of service will make it possible to keep electricity flowing to customers even when long-distance lines are turned off in fire-prone areas. The logic for microgrids has been around for a long time. See Hanna et al. 2019. Wildfires have altered the potential value from these deployments, although whether those values are realized is entirely context dependent and highly uncertain at present. Storage is pivotal to the idea, and like CPUC's battery procurement program, an experimentalist approach to deployment and learning through context is now underway. See St. John 2021.

69. For example, there is enormous variation in the reliability of the existing grid service. In remote communities served by a single power feeder line, grid reliability is often relatively low—especially if that line runs through a mountainous brushy area, and thus is periodically disconnected when it is windy and dry to avoid touching off wildfires. By contrast, communities served by many interconnected feeder lines will have more reliable grid service because the system overall is less vulnerable to the loss of any single component. In principle, these differences in grid configuration will lead to big differences in the value gained to a community from installing microgrids that keep operating even when the larger grid fails. While that variation in value is appreciated in theory, it is not yet possible to measure reliably how different configurations of macro- and microgrids affect the soundness of different microgrid investments.

70. See especially California Energy Storage Alliance, https://storagealliance.org/. See also California Solar and Storage Association, https://calssa.org/. At the federal level, the industry is also increasingly well organized. See Energy Storage Association, https://energystorage.org/.

71. Pyper 2017.

72. "About REDD+," 2021.

73. The laws, administrative measures, and statistics under discussion here assume different definitions of the Amazon. The most comprehensive is the Legal Amazon, an area of more than five million square kilometers, including the Brazilian states of Acre, Amapá, Amazonas, Maranhão, Mato Grosso, Pará, Rondônia, Roraima, and Tocantins, established by Brazilian federal law in 1953 in order to fix the scope of regional development and environmental policies. Some 25 percent of this area was originally covered by the Cerrado or savanna and other nonforest vegetation. The Amazon biome or habitat is contained within the the Legal Amazon except for a tiny area in the state of Maranhão. We refer to all of these indiscriminately as the Amazon. West and Fearnside 2021, 2.

74. Brandão et al. 2020.

75. Lucas and Warren 2013.

76. For a case study of these developments in a forest district in Indonesian Borneo, see Gecko Project 2017.

77. Campbell 2015.

78. Campbell 2015, 60.

79. Schwartzman 1991, 399.

80. For a thoughtful development of this view, well beyond the limits of the "primitive," however understood, see Western 2020.

81. Schwartzman et al. 2010, 277. Reference is to Alcorn 2005.

82. Arima et al. 2014.

83. Hochstetler and Keck 2007.

84. Locke 2013; Bartley 2018; Overdevest and Zeitlin 2018; Gereffi, Regini, and Sabel 2014.

85. Hochstetler and Keck 2007.

86. Riofrancos 2020.

87. Schwartzman et al. 2010.

88. Brandão et al. 2020, 4; Nepstad et al. 2014, 1118.

89. Jackson 2015.

90. Thaler, Viana, and Toni 2019.

91. West and Fearnside 2021. But for early signs that the registration program is coming into wider use and reducing deforestation rates among smallholder registrants, see Lipscomb and Prabakaran 2020.

92. Gibbs et al. 2016.

93. Nepstad et al. 2014, 1118.

94. "Based on a panel of Amazon municipalities from 2002 through 2009 . . . deforestation responded to agricultural output prices. After controlling for price effects . . . conservation policies

implemented beginning in 2004 and 2008 significantly contributed to the curbing of deforestation." Assunção, Gandour, and Rocha 2015. See also Assunção and Rocha 2019; Arima et al. 2014.

95. Gibbs et al. 2016; Heilmayr 2019; Picoli et al. 2020.

96. Assunção et al. 2020.

97. "About REDD+," 2021.

98. Since a low in the early 2010s, deforestation has more than doubled by 2020, but is still less than half the peak levels of the early 2000s. See BBC 2020.

99. Gomes, Perz, and Vadjunec 2012.

100. Hoelle 2014, 2015.

101. For details, see Kröger 2020; Hofmeister 2020. "The total for [2019] had already reached 74.5 square kilometers (28.7 square miles), three times more than the average in each of the past five years. (The latter figure was itself twice as high as registered before 2013, when the annual deforestation rate didn't exceed 10 km^2 [3.8 mi^2].)"

102. "Half of protected areas in the Brazilian Amazon have no approved management plan and 45 percent have no management council." Butler 2011. For a case study of disorganization, see Freitas, Rodrigues, and Silva 2016.

103. "Isso porque a maioria das iniciativas aplicadas era de natureza regulatória ou meramente repressiva, incapaz de alterar a dinâmica das atividades produtivas vinculadas ao desmatamento e, ao mesmo tempo, fomentar uma nova base eco- nômica sustentável na região. Além disso, a sociedade local, fortemente impactada com essas medidas, não percebia como vantajosa a questão ambiental e, por isso, não se engajava no processo de ordenamento. Foi nesse contexto que surgiram iniciativas locais ou regionais de combate ao desmatamento com potencial enorme de sustentar e aprofundar os ganhos até então obtidos, como é o caso do Programa Municípios Verdes no Estado do Pará." Whately, Campanili, and Veríssimo 2013.

104. Schmink et al. 2019.

105. For a comparison of Paragominas and São Felix, see Brandão et al. 2020. For a full account of São Felix, see Schmink et al. 2019.

106. Varns et al. 2018.

107. Schmink et al. 2019.

108. Thaler, Viana, and Toni 2019.

109. Brandão et al. 2020.

110. In the increasing weight of smallholder deforestation after the command-and-control reforms, see F. Seymour and Harris 2019; Schneider et al. 2015. See also Schielein and Börner 2018.

111. Vargas 2020.

112. Wallace, Gomes, and Cooper 2018.

113. Kröger 2020.

114. "Povos Indígenas do Rio Negro Planejam o Futuro que Desejam" 2017; Navarro 2020.

115. Which is not to say, of course, that even with regard to the regulation of larger property, the measures have been a panacea. For ways to view them, see Branford and Torres 2017.

116. Schmink et al. 2019.

117. For an account of the widespread embrace of no-till agriculture among US farmers in the Midwest and Great Plains concerned about soil erosion but reluctant to discuss climate change, see Horn 2017.

118. Santiago, Caviglia-Harris, and Pereira de Rezende 2018.

119. Skidmore 2020.

120. "Smallholder Crucial to Preserving the Amazon Rainforest" 2019; de Paiva Serôa da Motta et al. 2018.

121. For the potential uses of latex and its by-products, see Jaramillo-Giraldo et al. 2017.

122. Coslovsky 2021.

123. Coslovsky 2014.

124. Buainain, Lanna, and Navarro 2020; Correa and Alkmin Junqueira Schmidt 2014.

125. Guanziroli et al. 2019.

126. For details, see Guanziroli et al. 2019; Rada, Helfand, and Magalhães 2019. The figures are from 2006. A census of agricultural holdings was scheduled for 2016, but it was postponed because of domestic turmoil. Between 1985 and 2006, the productivity of land increased with the size of farms, reflecting economies of scale in the use of that factor. But the size distribution of the total factor productivity, taking into consideration the productivity of labor and capital as well, was U shaped, indicating that both small and large farms can combine endowments with increasing efficiency. Some of the efficient small producers supply oranges, chickens, or pigs to large food processors, which provide them with key inputs and prescribe production methods that must be rigidly followed. Such industrial supply chains are as a rule environmentally dirty. Other efficient smallholders grow fresh produce by cleaner, more skill-intensive methods. Sustainable, economically viable smallholder ranching resembles the latter.

127. Pacheco et al. 2018; Schoneveld et al. 2019; Dharmawan et al. 2021.

128. Nature Conservancy 2018; Seymour, Aurora, and Arif 2020; Stickler et al. 2018; Buchanan et al. 2019; Fishman, Oliveira, and Gamble 2017; Brandão et al. 2020.

129. For an incisive discussion of the choices that rural society in Brazil faces, see Navarro 2020.

Chapter 6. International Cooperation beyond Paris

1. Victor and Jones 2018; Victor 2019.

2. The Swedish strategy involves a partnership between Swedish steel firms and an energy supplier, Vattenfall, that has a specialty in hydrogen. Hoikkala and Starn 2020.

3. The value of scale is greater for small countries, of course, but even for large countries cooperation can expand the range of the technological ideas, policy strategies, and business models tested. Other methods include capturing carbon dioxide during hydrogen production and the use of electricity to transform, chemically, iron into steel without carbon dioxide as a by-product. "Accelerating the Energy Transition" 2016; Victor, Geels, and Sharpe 2019.

4. Dolphin, Pollitt, and Newbery 2020; Cullenward and Victor 2020.

5. Dolphin, Pollitt, and Newbery 2020.

6. Abnett 2021.

7. In one of these areas, forestry, there is a formal link back into the Paris Agreement through the Reducing Emissions from Deforestation and Forest Degradation system discussed in chapter 5.

8. Raustiala and Victor 2004; Alter and Meunier 2009; Oberthür and Stokke 2011; Biermann et al. 2009. Related ideas are advanced around the governance of problems that don't implicate international cooperation. See Ostrom 2010; Hawkins et al. 2006.

9. Sabel and Victor 2015; Sabel and Victor 2016.

10. Sabel and Victor 2015, 2016; Victor 2015.

11. See, for example, Morales 2015; Victor 2015; Falkner 2016; Bodansky 2016.

12. See paragraph 24 and subsequent decisions in UNFCCC 2018.

13. Victor 2015; Hale 2017.

14. Hale 2016. Formally, some assessment of these subnational efforts does exist, organized partly in response to the Marrakech Partnership for Global Climate Action by the High-Level Champions in 2016. See, for example, UNFCCC 2020. For an appraisal of the efforts at assessment, see Hale et al. 2021.

15. Of course, in the case of Montreal, national commitments were codified in a different way. All members subscribed to common commitments in the form of targets and timetables; national and sectoral differentiation arose through the effort to cut emissions, which often revealed places

of difficulty such as essential uses that required exemption (see chapter 2). This kind of differentiation of common commitments through local, reviewable experience is used in many different international agreements and institutions—for example, through Article IV of the International Monetary Fund. See Chayes and Chayes 1991. In Paris, by contrast, the differentiation is built into the pledges—the NDCs—along with the differentiation revealed through a national effort.

16. Not only was Paris itself negotiated under consensus rules, but the negotiations elaborating most of the key procedures (e.g., the detailed reporting via the NDCs and facilitative multilateral consultative process review) were subject to the same procedural rules as well. Those negotiations were launched right after the Paris Agreement was established. Hale 2016; Rajamani and Bodansky 2019.

17. At this writing, the rulebook negotiations on Article 6, which implicate international emissions trading and other topics, are still incomplete.

18. The role of consensus remains on display, at this writing, around Article 6, which many countries see as the mechanism through which emission trading might occur. There the controversies continue to simmer—partly because international emission trading remains an abstract possibility with few tangible benefits—and thus the rulebook has not been finalized. Stavins 2019; Cullenward and Victor 2020, chapter 6.

19. What can be observed so far is the evolution of the NDCs, which is not encouraging. We note, however, that some language of Article 13 of the Paris Agreement (as opposed to the narrow rules of Article 4) read together with the "country reports" produced under the UNFCCC, might be construed as enabling NDCs detailed enough to serve as the basis for experimentalist review.

20. "Update of Norway's Nationally Determined Contribution" 2020.

21. OGJ editors 2020. For a discussion of Norwegian contributions to the Amazon Fund, see chapter 5.

22. Formally, the Global Climate Action predates Paris and was originally known as the Nonstate Actor Zone for Climate Action portal. It emerged, in part, from the massive attention to subnational and corporate action on climate change displayed at the UN General Assembly in September 2014. At the next COP following that event, in Lima, governments agreed on a process that became known as the Lima-Paris Action Agenda and from there the NAZCA portal.

23. In Glasgow the UN Secretary General announced plans for a new program that would review subnational policies—a good idea in principle, although one that can't be made effective if it follows UN consensus rules.

24. For, more generally, the differentiation of commitments and expectations in the Paris process, see Voigt and Ferreira 2016.

25. Substantial resources are devoted to these, although the agendas are so broad that the procedures are quite far from the detailed postexperiment review that characterizes how the Montreal TOCs work. For a summary of the meetings in 2019, see UNFCCC 2019.

26. The climate change secretariat, which organized the Technical Expert Meetings (TEMs), was established first and foremost to facilitate negotiations, not to delve into actual substantive issues. The secretariat would basically line up a long list of speakers—parties and nonparties that, over a TEM, would talk about what they were doing and what worked (and didn't work), but this involved long panel discussions in empty rooms. What could have evolved into something more experimental—a framing of possible new directions and advice to the parties about ways to adjust top-level commitments—was further encumbered by the fact that many governments sent diplomats to the TEMs, as opposed to actual experts or regulatory decision-makers grappling with the substance.

27. Adjacent to the Paris Agreement, however, are funding mechanisms that are, to varying degrees, gathering this information and, in their own diverse ways, operating more along the lines of experimentalism. For example, the long-standing Global Environment Facility includes some of

this experimental learning, as do some of the multilateral development banks that operate in the same space. The Green Climate Fund, if not perennially beset with political difficulties and contestation, could evolve in this way. (That "if" is important, for an intergovernmental body designed like the Green Climate Fund might be incapable of operating in a nimble fashion.) Future research on the funding regimes that are adjacent to Paris—most of which have their own decision-making systems—should look at and explain the degree of experimentalism in these kinds of institutions. Experimentalism and learning in international funding mechanisms may prove to be one of the most important topics in international cooperation on climate change.

28. In advancing this argument, we resonate with the assertions made in Hale 2020; Young 2017. Where we differ is in our view of how the machinery that uses these goals is experimentalist in its functioning.

29. Of course there are many hybrid or intermediate cases, such as soybeans—a commodity—grown from seeds optimized to highly local conditions, or a standard piece of mining equipment customized to local geological conditions and regulatory requirements.

30. In the real world, technology, business, and social science will involve blends of these different functions; a revolution in electric vehicles, for example, involves a mixture of the first functions (e.g., battery and vehicle manufacture) and local contextualization (e.g., local charging networks). We articulate here the bookends to allow a focus on the functional attributes.

31. Emissions from electricity are harder to categorize. On the one hand, many key electric technologies are manufactured to global standards and sold in global markets (e.g., refrigerators and solar panels). On the other, the core challenge of decarbonizing electricity and expanding electric networks is a matter of contextualization, as we show in chapter 5. On the breakdowns of emissions by sectors, see Victor et al. 2014.

32. International Energy Agency 2020a, 2020b.

33. A few sector initiatives are emerging—for instance, the Oil and Gas Climate Initiative, a group of firms that aim to advance carbon capture and storage as well tighten controls on leaks of methane, a potent warming gas. The initiative invests in new technologies (and works with members that invest in those technologies as well). As is typical of such collaborations, there is frequent stocktaking (e.g., annually). In the world of Paris, by contrast, the stocktaking process, as we noted, runs every five years; controlled by governments, it is slow and broad, and likely to be encumbered by rules established through intergovernmental consensus.

34. Buchanan and Tullock 1962; Nordhaus 2015; Victor 2011.

35. This is a familiar problem in industry groups. See, for example, Rees 1996. It is also a problem internationally, often addressed in the past through hegemonic cooperation. For a look at this problem applied to climate change, see Victor, Geels, and Sharpe 2019.

36. On aviation, see, for example, Carroll 2021.

37. Maersk has pledged net-zero emissions by 2050, which means nearly immediate changes in its fleet since a typical ship lasts twenty to twenty-five years. See "IMO Agrees to CO2 Emissions Target" 2018.

38. As a practical example, see Martin 2020. For the theory, see Kahler 1992; Buchanan and Tullock 1962; Victor 2017. For the theory applied to the observable record, see Downs, Rocke, and Barsoom 1996.

39. The North Sea and Montreal Protocol cases—problem-solving through fine-grained sectoral approaches—are discussed in chapter 2.

40. Parson 2003.

41. Energy Transitions Commission 2020b; "Accelerating the Energy Transition" 2016; Victor, Geels, and Sharpe 2019.

42. A weakness for ARPA-E is that the DOE does not have reliable programs to help pick up these promising combinations of technological innovation and help them cross the "valley of

death." ARPA-E feeds them, gets them to the valley, and many then die for lack of demonstration support. For more on this, see Buchanan and Tullock 1962. And for proposals aimed at boosting energy innovation that are cognizant of this problem and thus aim for a huge ramping up of demonstration support, see, Sivaram, Friedmann, and Cunliff 2020; National Academies of Sciences, Engineering, and Medicine 2021a, 2021b.

43. National Academies of Sciences, Engineering, and Medicine 2021a; Larson et al. 2020; Sustainable Development Solutions Network 2020; Clarke et al. 2014.

44. In the lingo of deep decarbonization, these sectors are often called "hard to abate" precisely because the solution that is a likely winner in other sectors, namely electrification, is not plausibly available for these sectors. Energy Transitions Commission 2018.

45. At this writing, the European effort is the most credible, but complementary efforts in other countries are also emerging, notably in Japan and the United States. A degree of competition in creating what is widely seen as an industry of the future has spurred investment even as global supply chains for key hydrogen technologies (e.g., electrolyzers) will bring down the costs with scale just as occurred with solar power. On the solar experience, see Nemet 2019.

46. This problem of keeping a consensus club together exists not just for investing in technologies of the future but also for initiatives that help speed the exit of the old guard. For example, the European Bank for Reconstruction and Development has operated a pioneering program aimed at showing how to help countries move beyond coal by shutting down old facilities while taking care of affected communities—a program that where successful, would help open up frontiers for noncoal sources of energy. When the Trump administration came to power, it worked to block such programs, as the United States was a founding member of the European Bank for Reconstruction and Development, and only because other pioneering members of the bank and secretariat found ways to navigate around US intransigence was it possible for the beyond-coal campaign to continue.

47. Energy Transitions Commission 2020a.

48. The one notable exception to this statement is in China, where there are substantial frontier-pushing investments in climate-related areas (e.g., advanced nuclear power, electric vehicles, solar, and wind) that have historically had little to do with climate change, and are more firmly anchored in other overlapping concerns such as energy security, industrial development, and local air pollution. Over time, partly due to export opportunities, the Chinese investment has become more climate focused.

49. For example, in the oil and gas industry, the consortium invests in decarbonization technologies, including carbon capture and storage. Oil and Gas Climate Initiative 2019. And there is a parallel initiative focused just on methane—which began with detailed reporting about methane control methods, and then adjusted the central rules to require more information and action about the absolute volumes of methane emissions. UNEP 2020.

50. National Academies of Sciences, Engineering, and Medicine 2021b, especially chapter 3.

51. For more on the Regulatory Assistance Project, see https://raponline.org/.

52. Skjærseth 1998.

53. Ogden and Marano 2016; G20 Working Group on Energy and Commodities 2012.

54. For the best overview, see Ogden and Marano 2016. For the self-reviews and peer reviews, see China et al. 2016; Germany et al. 2016. For the Chinese and US self-reports, see United States 2015; China 2015. The US presidential election in 2016, just a month after the reviews were published, helped put the bilateral relationship into a deeper freeze, from which it has not recovered—a reminder that volunteer-based activities can help to demonstrate models and catalyze broader efforts, but are fickle when the volunteers focus on other tasks. Volunteerism must be followed by institutionalization. At this writing, it is not clear that the United States and China would be the ideal starting partners for this kind of mutual review because both countries are increasingly wary of the other; a bigger tent, such as a quadrilateral approach in the Pacific or transatlantic

US-EU-UK trilateral system, might be more effective and less dependent on the already-stressed bilateral US-China relationship.

55. Azevedo et al. 2020b; Rabe 2004; America's Pledge Initiative on Climate Change 2020.

56. Azevedo et al. 2020b.

57. C40 (with ninety-seven affiliated cities at this writing) has created a series of forums where city officials can compare notes on topics like land use planning and building efficiency; it has also created some standards for measuring emissions, with minimal impact. See C40 Cities, https://www.c40.org/research. In a few cities (eight at this writing), C40 sends advisers to consult on best practices. Local Governments for Sustainability (ICLEI), bigger and broader (twenty-five hundred cities, towns, and regions at this writing), is a little more active in assessing city-level performance, mainly for the purposes of offering thin case studies on success stories on topics such as efficient street lighting, fundraising, and urban planning—but there is no systematic performance review of the organization's members and no requirement for review as a condition of membership.

58. Chayes and Chayes 1998.

59. Stone 2018.

60. Mallet and Khalaf 2020.

61. In particular, the Biden plan would "condition future trade agreements on partners' commitments to meet their enhanced Paris climate targets." "The Biden Plan" 2020. It is important not to overstate the credibility of this position, however, since at this writing (mid-2021), the Biden administration, now in power, also realizes that the growing momentum for trade measures in Europe is partly intended to punish the United States if it drags its feet. See Geman 2021.

62. Because Paris is a visible rallying point for action, it is mostly helpful in focusing political energy, but in some settings it has the opposite effect—such as with the Far Right in the United States, where the visibility of Paris made it easier to mobilize opposition, ultimately leading to the US announcement of withdrawal under Trump. That opposition and the conspicuous withdrawal, however, created a counterresponse (e.g., the "We Are Still In" movement) that induced further effort within the United States in states and localities as well as among financial regulators and firms. All told, that subnational action—itself inspired by Paris as the conscience of the world—is probably proving more durable than many federal actions that have been easy to reverse. Victor, Frank, and Gesick 2020.

63. Indeed, at this writing a "carbon border adjustment mechanism" is advancing rapidly in Europe with possible countermeasures also contemplated in the United States.

64. Bacchus 2018; WTO 2001a, 2001b.

65. The procedures for information provision and review are designed to make discrimination and tailoring to country circumstances difficult, except where the country itself might invite that. Some of the funding mechanisms active on climate-related topics are investing in local capacity building similar to the MLF-type funding mechanisms, among them the World Bank, European Bank for Reconstruction and Development, Inter-American Development Bank, and Asian Development Bank. A full analysis of the constellation of funding mechanisms relevant to climate is outside the scope of this book, but we think it's a high-priority topic for future research.

66. Chayes and Chayes 1998.

67. Notably, see UNFCCC 2015, Article 15. This article is reminiscent of the "multilateral consultative process" of the UNFCCC (see UNFCCC 2015, Article 13), and offers a "facilitative" approach to implementation that in principle, is "expert-based" and attentive to "respective national capabilities and circumstances of the Parties." A pioneer country or countries could demonstrate facilitated implementation along the lines we suggest and point to consistency with Article 15. We remain skeptical that a formally established Article 15 process—which would need to follow rules and procedures established through a mechanism like the rulebook—could actually function the way we suggest here.

68. Bagehot 1867.

Chapter 7. Piecing Together a More Accountable Globalization

1. This is a paraphrase of the aim of the European Union's Carbon Border Adjustment Mechanism, which aims to induce all countries to adopt market-based mechanisms for controlling pollution and then adjust prices at the borders to equalize the impact of these policies on industry. While simple in theory, the market mechanisms proposed for comparing abatement levels across borders are similar to those introduced in the CDM and suffer, despite much subsequent refinement, from the same defects. The mix of national and local policies that affect the behavior of firms is impossible to observe from the outside, as are the counterfactuals, and thus identifying the border tariff equivalents will be unworkable. For more on the Carbon Border Adjustment Mechanism, see Abnett and Twidale 2021. For a more protectionist variant—a US proposal—see Coons and Peters 2021.

2. For an insightful review of the technocrats' overconfidence in their predictions of the benefits of trade, obstinate lack of interest in evidence to the contrary, doubts about the ability of government to identify those harmed by its policies, but the conviction that compensation, if any, should be through the tax-and-transfer system, not the regulation of the economy, see Raskolnikov 2021.

3. Autor, Dorn, and Hanson 2013.

4. Case and Deaton 2015, 2020.

5. Rodrik 2011.

6. Rodrik 2011.

7. This continues to be the standard view among economists. For a recent restatement, see Nordhaus 2021.

8. Jennison 2013. For background developments in the European Union, see Soekkha 1997. For an overview of the agreement as an example of regulatory equivalence, see Sabel 2019. Like many such collaborations between national regulators outside preferential trade agreements, the EU-US Bilateral Aviation Safety Agreement was formalized as an executive agreement. Such agreements have their own legitimacy deficits. See Hathaway, Bradley, and Goldsmith 2020, 629. The implementation of joint regulatory programs across national borders can also be difficult because of the same kinds of organizational problems and conflicts of interest that make domestic reform difficult. For a sobering assessment of US-Canadian cooperation on food safety, see Kerr and Hobbs 2021.

9. On Australia and New Zealand as leaders for the adoption of Hazard Analysis Critical Control Point food safety standards starting in the 1990s, see Ropkins and Beck 2000.

10. This draws on Hoekman and Sabel 2019. For the OPA concept as it is emerging in trade talks, see, for example, Buchanan 2020.

11. Just how closed PTAs can be is demonstrated by Great Britain's continuing difficulties in finding any commercially feasible alternatives to the continuing, deep engagement with EU regulators—at odds with the reassertion of national sovereignty that motivated Brexit. The European Union's accession and neighborhood policies are the exceptions that prove the rule as it focuses on building the capacity of candidate members to meet regulatory requirements. For a nuanced discussion, see Lavenex 2015.

12. Hoekman 2016; Hoekman and Sabel 2019.

13. For an account of how the imposition of such a choice on the peoples of eastern Europe after the fall of the Berlin Wall allowed populist parties to present themselves as champions of national values against predatory cosmopolitan elites, see Krastev and Holmes 2019.

REFERENCES

Abbott, Kenneth W., Philipp Genschel, Duncan Snidal, and Bernhard Zangl. 2012. "Orchestration: Global Governance through Intermediaries." https://doi.org/10.2139/ssrn.2125452.

Abnett, Kate. 2021. "EU Sees Carbon Border Levy as 'Matter of Survival' for Industry." *Reuters*, January 18. https://www.reuters.com/article/uk-climate-change-eu-carbon/eu-sees-carbon-border-levy-as-matter-of-survival-for-industry-idUKKBN29N1O0.

Abnett, Kate, and Susanna Twidale. 2021. "EU Proposes World's First Carbon Border Tax for Some Imports." *Reuters*, July 14. https://www.reuters.com/business/sustainable-business/eu-proposes-worlds-first-carbon-border-tax-some-imports-2021-07-14/.

"About REDD+" 2021. UN-REDD Programme, April 12. https://www.unredd.net/about/what-is-redd-plus.html.

"AB-2514 Energy Storage Systems." 2010. California Legislative Information. https://leginfo.legislature.ca.gov/faces/billNavClient.xhtml?bill_id=200920100AB2514.

"Accelerating the Energy Transition: Cost or Opportunity?" 2016. New York: McKinsey & Company. https://www.mckinsey.com/featured-insights/europe/accelerating-the-energy-transition-cost-or-opportunity.

Ackerman, Bruce A., and Richard B. Stewart. 1984. "Reforming Environmental Law." *Stanford Law Review* 37:1333–65.

Administrative Procedure Act. 1946. 5 U.S.C. ch. 5 §553.

Alcorn, Janice. 2005. "Dances around the Fire: Conservation Organizations and Community-Based Natural Resource Management." In *Communities and Conservation: Histories and Politics of Community-Based Natural Resource Management*, edited by J. Peter Brosius, Anna Lowenhaupt Tsing, and Charles Zerner, 37–68. Globalization and the Environment. Lanham, MD: Rowman and Littlefield.

Alothman, Mohammad, Sean A. Hogan, Deirdre Hennessy, Pat Dillon, Kieran N. Kilcawley, Michael O'Donovan, John Tobin, Mark A. Fenelon, and Tom F. O'Callaghan. 2019. "The 'Grass-Fed' Milk Story: Understanding the Impact of Pasture Feeding on the Composition and Quality of Bovine Milk." *Foods* 8 (8): 350. https://doi.org/10.3390/foods8080350.

Alter, Karen J., and Sophie Meunier. 2009. "The Politics of International Regime Complexity." *Perspectives on Politics* 7 (1): 13–24. https://scholar.princeton.edu/sites/default/files/altermeunierperspectives_0.pdf.

American Bar Association. 1993. "ABA Recommendation 120C, 118–2 Ann. Rep. A.B.A. 57–58." Chicago: American Bar Association.

American Electric Power. 2017. "AEP Employees Win EPRI Technology Transfer Awards." https://aepretirees.com/2017/02/14/aep-employees-win-epri-technology-transfer-awards/.

America's Pledge Initiative on Climate Change. 2020. "Delivering on America's Pledge: Achieving Climate Progress in 2020." New York: Bloomberg Philanthropies with Rocky Mountain

Institute, University of Maryland Center for Global Sustainability, and World Resources Institute. https://assets.bbhub.io/dotorg/sites/28/2020/09/Delivering-on-Americas-Pledge.pdf.

Andersen, Stephen O., and K. Madhava Sarma. 2004. *Protecting the Ozone Layer: The United Nations History*. London: Earthscan.

Andersen, Stephen O., Nancy J. Sherman, Suely Carvalho, and Marco Gonzalez. 2018. "The Global Search and Commercialization of Alternatives and Substitutes for Ozone-Depleting Substances." *Comptes Rendus Geoscience* 350 (7): 410–24. https://doi.org/10.1016/j.crte.2018.07.010.

"Annual Levelized Cost of Energy Analysis (LCOE 14.0)." 2020. Lazard. https://www.lazard.com/perspective/levelized-cost-of-energy-and-levelized-cost-of-storage-2020/.

Arima, Eugênio Y., Paulo Barreto, Elis Araújo, and Britaldo Soares-Filho. 2014. "Public Policies Can Reduce Tropical Deforestation: Lessons and Challenges from Brazil." *Land Use Policy* 41 (November): 465–73. https://doi.org/10.1016/j.landusepol.2014.06.026.

"ASSAP—Farming for Water Quality." n.d. Teagasc. https://www.teagasc.ie/environment/water-quality/farming-for-water-quality---assap/.

Association of American Railroads. 2011. "The Impact of the Staggers Rail Act of 1980." Washington, DC: Association of American Railroads.

Assunção, Juliano, Clarissa Gandour, Romero Rocha, and Rudi Rocha. 2020. "The Effect of Rural Credit on Deforestation: Evidence from the Brazilian Amazon." *Economic Journal* 130 (626): 290–330. https://doi.org/10.1093/ej/uez060.

Assunção, Juliano, Clarissa Gandour, and Rudi Rocha. 2015. "Deforestation Slowdown in the Brazilian Amazon: Prices or Policies?" *Environment and Development Economics* 20 (6): 697–722. https://doi.org/10.1017/S1355770X15000078.

Assunção, Juliano, and Romero Rocha. 2019. "Getting Greener by Going Black: The Effect of Blacklisting Municipalities on Amazon Deforestation." *Environment and Development Economics* 24 (2): 115–37. https://doi.org/10.1017/S1355770X18000499.

Ausubel, Jesse H., Iddo K. Wernick, and Paul E. Waggoner. 2013. "Peak Farmland and the Prospect for Land Sparing." *Population and Development Review* 38 (January): 221–42.

Autor, David H., David Dorn, and Gordon H. Hanson. 2013. "The China Syndrome: Local Labor Market Effects of Import Competition in the United States." *American Economic Review* 103 (6): 2121–68. https://doi.org/10.1257/aer.103.6.2121.

Ayres, Ian, and Robert Gertner. 1989. "Filling Gaps in Incomplete Contracts: An Economic Theory of Default Rules." *Yale Law Journal* 99 (1): 87–130.

Azevedo, Inês, Michael R. Davidson, Jesse D. Jenkins, Valerie J. Karplus, and David G. Victor. 2020a. "The Paths to Net Zero: How Technology Can Save the Planet." *Foreign Affairs* 99 (3): 18–27.

Azevedo, Inês, Sam Markolf, Mark Muro, and David G. Victor. 2020b. "Pledges and Progress: Steps toward Greenhouse Gas Emissions Reductions in the 100 Largest Cities across the United States." Washington, DC: Brookings Institution. https://www.brookings.edu/research/pledges-and-progress-steps-toward-greenhouse-gas-emissions-reductions-in-the-100-largest-cities-across-the-united-states/.

Azoulay, Pierre, Erica Fuchs, Anna P. Goldstein, and Michael Kearney. 2019. "Funding Breakthrough Research: Promises and Challenges of the 'ARPA Model.'" *Innovation Policy and the Economy* 19 (January): 69–96. https://doi.org/10.1086/699933.

Bacchus, James. 2018. *The Willing World: Shaping and Sharing a Sustainable Global Prosperity*. Cambridge: Cambridge University Press. https://doi.org/10.1017/9781108552417.

Bagehot, Walter. 1867. *The English Constitution*. London: Chapman and Hall.

Barker, Brent. 2012. "Born in a Blackout." *EPRI Journal* (Summer). http://mydocs.epri.com/docs/CorporateDocuments/EPRI_Journal/2012-Summer/1025689_Blackout.pdf.

Barrett, Scott. 1991. "Economic Analysis of International Environmental Agreements: Lessons for a Global Warming Treaty." In *Responding to Climate Change: Selected Economic Issues*, edited by the Organization for Economic Cooperation and Development. Paris: Organization for Economic Cooperation and Development.

———. 2005. "The Montreal Protocol." In *Environment and Statecraft: The Strategy of Environmental Treaty-Making*, edited by Scott Barrett, 221–53. Oxford: Oxford University Press. https://doi.org/10.1093/0199286094.001.0001.

Bartley, Tim. 2018. *Rules without Rights: Land, Labor, and Private Authority in the Global Economy*. Transformations in Governance. Oxford: Oxford University Press.

Battery Technology Advisory Panel (BTAP). 2000. "Advanced Batteries for Electric Vehicles: An Assessment of Performance, Cost and Availability." Sacramento: California Air Resources Board.

Baucus, Max S. 1988. "Conference on Substitutes and Alternatives for CFCs (Chlorofluorocarbons) and Halons (January 14, 1988)." *Max S. Baucus Speeches and Remarks* 400 (January). https://scholarworks.umt.edu/cgi/viewcontent.cgi?article=1399&context=baucus_speeches&usg=AOvVaw22OrBBXHgq4fhCBbQdrVLt.

BBC. 2020. "Brazil's Amazon: Deforestation 'Surges to 12-Year High.'" *BBC*, November 30, 2020. https://www.bbc.com/news/world-latin-america-55130304.

Bedsworth, Louise Wells, and Margaret R. Taylor. 2007. "Learning from California's Zero-Emission Vehicle Program." *California Economic Policy* 3 (4): 1–19.

Benedick, Richard E. 1991. *Ozone Diplomacy: New Directions in Safeguarding the Planet*. Cambridge, MA: Harvard University Press.

———. 1998. *Ozone Diplomacy: New Directions in Safeguarding the Planet*. Cambridge, MA: Harvard University Press.

Ben-Shahar, Omri, and James J. White. 2006. "Boilerplate and Economic Power in Auto Manufacturing Contracts." *Michigan Law Review* 104 (5): 953–82.

"The Biden Plan for a Clean Energy Revolution and Environmental Justice." 2020. https://joebiden.com/climate-plan/.

Biermann, Frank. 1996. "Financing Environmental Policies in the South: An Analysis of the Multilateral Ozone Fund and the Concept of Full Incremental Costs, WZB Discussion Paper, No. FS II 96–406." *Wissenschaftszentrum Berlin für Sozialforschung (WZB)*. http://bibliothek.wz-berlin.de/pdf/1996/ii96-406.pdf.

Biermann, Frank, and Steffen Bauer, eds. 2005. *A World Environment Organization: Solution or Threat for Effective International Environmental Governance?* Global Environmental Governance Series. Aldershot, UK: Ashgate.

Biermann, Frank, Philipp Pattberg, Harro van Asselt, and Fariborz Zelli. 2009. "The Fragmentation of Global Governance Architectures: A Framework for Analysis." *Global Environmental Politics* 9 (4): 14–40. https://doi.org/10.1162/glep.2009.9.4.14.

Bodansky, Daniel. 1993. "The United Nations Framework Convention on Climate Change: A Commentary." *Yale Journal of International Law* 18 (2): 451–558.

———. 2001. "The History of the Global Climate Change Regime." In *International Relations and Global Climate Change*, edited by Urs Luterbacher and Detlef F. Sprinz, 23–40. Cambridge, MA: MIT Press.

———. 2016. "The Paris Climate Change Agreement: A New Hope?" *American Journal of International Law* 110 (2): 288–319.

Bonvillian, William, Richard Van Atta, and Patrick Windham. 2019. *The DARPA Model for Transformative Technologies: Perspectives on the U.S. Defense Advanced Research Projects Agency*. OpenBook Publishers. http://doi.org/10.11647/OBP.0184.

Bord Bia. 2016. "Performance and Prospects 2015–2016, January 2016." https://www.bordbia.ie/industry/insights/publications/performance-and-prospects-2015-2016/.

Boyd, William. 2021. "The Poverty of Theory: Public Problems, Instrument Choice, and the Climate Emergency." *Columbia Journal of Environmental Law* 46 (2): 399–487. https://doi.org/10.52214/cjel.v46i2.8401.

Boyle, Richard, Joanna O'Riordan, Fergal O'Leary, and Laura Shannon. 2021. *Using an Experimental Governance Lens to Examine Governance of the River Basin Management Plan for Ireland 2018–2021.* Johnstown Castle, Ireland: Environmental Protection Agency.

Bradford, Anu. 2020. *The Brussels Effect: How the European Union Rules the World.* Oxford: Oxford University Press. https://doi.org/10.1093/oso/9780190088583.001.0001.

Brandão, Frederico, Marie-Gabrielle Piketty, René Poccard-Chapuis, Brenda Brito, Pablo Pacheco, Edenise Garcia, Amy E. Duchelle, Isabel Drigo, and Jacqueline Carvalho Peçanha. 2020. "Lessons for Jurisdictional Approaches from Municipal-Level Initiatives to Halt Deforestation in the Brazilian Amazon." *Frontiers in Forests and Global Change* 3 (August): 96. https://doi.org/10.3389/ffgc.2020.00096.

Branford, Sue, and Maurício Torres. 2017. "Amazon Soy Moratorium: Defeating Deforestation or Greenwash Diversion?" Mongabay, March 8. https://news.mongabay.com/2017/03/amazon-soy-moratorium-defeating-deforestation-or-greenwash-diversion/.

Brusoni, Stefano, and Andrea Prencipe. 2001. "Unpacking the Black Box of Modularity: Technologies, Products and Organizations." *Industrial and Corporate Change* 10 (1): 179–205. https://doi.org/10.1093/icc/10.1.179.

Buainain, Antônio Márcio, Rodrigo Lanna, and Zander Navarro. 2020. *Agricultural Development in Brazil: The Rise of a Global Agro-Food Power.* London: Routledge.

Buchanan, James, and Gordon Tullock. 1962. *The Calculus of Consent: Logical Foundations of Constitutional Democracy.* Ann Arbor: University of Michigan Press.

Buchanan, John, Joanna Durbin, Dave McLaughlin, Lesley McLaughlin, Katie Thomason, and Melissa Thomas. 2019. "Exploring the Reality of the Jurisdictional Approach as a Tool to Achieve Sustainability Commitments in Palm Oil and Soy Supply Chains." Conservation International. March. https://www.conservation.org/docs/default-source/publication-pdfs/jurisdictional_approach_full_report_march2019_published.pdf.

Buchanan, Kelly. 2020. "Chile; New Zealand; Singapore: Digital Trade Agreement Signed." June 17. https://www.loc.gov/item/global-legal-monitor/2020-06-17/chile-new-zealand-singapore-digital-trade-agreement-signed/.

Burgess, Edward. 2018. Interview with Rory O'Donnell.

Butler, Rhett A. 2011. "Protected Areas Cover 44% of the Brazilian Amazon." Mongabay, April 20. https://news.mongabay.com/2011/04/protected-areas-cover-44-of-the-brazilian-amazon/.

California Air Resources Board (CARB). 1994. "Staff Report: Low-Emission and Zero-Emission Vehicle Program Review." Sacramento: California Air Resources Board.

———. 1996. "Staff Report: Low-Emission and Zero-Emission Vehicle Program Review." Sacramento: California Air Resources Board.

———. 2012. "The Zero Emission Vehicle (ZEV) Regulation." Sacramento: California Air Resources Board.

California Energy Commission and Pamela Doughman. 2020. "Electric Program Investment Charge, 2019 Annual Report." CEC-500-2020-009. Sacramento: California Energy Commission. https://www.energy.ca.gov/publications/2020/electric-program-investment-charge-2019-annual-report.

California Independent System Operator (CAISO). 2011. "Business Requirements Specification—Regulation Energy Management (REM)—Non Generator Resource (NGR)." Folsom: California Independent System Operator.

———. 2014. “Draft Final Proposal—Energy Storage Interconnection.” Folsom: California Independent System Operator.

California Independent System Operator (CAISO), California Public Utilities Commission (CPUC), and California Energy Commission. 2021. “Final Root Cause Analysis: Mid-August 2020 Extreme Heat Wave.” http://www.caiso.com/Documents/Final-Root-Cause-Analysis-Mid-August-2020-Extreme-Heat-Wave.pdf.

California Public Utilities Commission (CPUC). 2008. “Decision Granting a Certificate of Public Convenience and Necessity for the Sunrise Powerlink Transmission Project.” Item 60b. San Francisco: California Public Utilities Commission. http://docs.cpuc.ca.gov/PUBLISHED/AGENDA_DECISION/95357.htm#P263_8685.

———. 2013. “Decision Adopting an Energy Storage Procurement Framework and Design Program.” Decision 13-10-040. San Francisco: California Public Utilities Commission. https://docs.cpuc.ca.gov/PublishedDocs/Published/G000/M079/K533/79533378.PDF.

———. 2014. “Decision Approving San Diego Gas & Electric Company, Pacific Gas and Electric Company, and Southern California Edison Company’s Storage Procurement Framework and Program Applications for the 2014 Biennial Procurement Period.” Decision 14-10-045. San Francisco: California Public Utilities Commission. http://docs.cpuc.ca.gov/PublishedDocs/Published/G000/M127/K426/127426247.PDF.

———. 2015. “Beyond 33% Renewables: Grid Integration Policy for a Low-Carbon Future.” CPUC Staff White Paper. California Public Utilities Commission Energy Division. https://www.cpuc.ca.gov/WorkArea/DownloadAsset.aspx?id=9141.

———. 2016. “Decision on Track 1 Issues.” Decision 16-01-032. San Francisco: California Public Utilities Commission. https://docs.cpuc.ca.gov/PublishedDocs/Published/G000/M158/K111/158111344.PDF.

———. 2018. “Decision on Multiple Use Application Issues.” Decision 18-01-003. San Francisco: California Public Utilities Commission. http://docs.cpuc.ca.gov/PublishedDocs/Published/G000/M206/K462/206462341.PDF.

Callon, Michel, Pierre Lascoumes, and Yannick Barthe. 2009. “Measured Action, or How to Decide without Making a Definitive Decision.” In *Acting in an Uncertain World: An Essay on Technical Democracy*, translated by Graham Burchell, 191–224. Inside Technology. Cambridge, MA: MIT Press.

Cames, Martin, Ralph O. Harthan, Jürg Füssler, Michael Lazarus, Carrie Lee, Peter Erickson, and Randall Spalding-Fecher. 2016. “How Additional Is the Clean Development Mechanism? Analysis of the Application of Current Tools and Proposed Alternatives. Study Prepared for DG CLIMA.” https://doi.org/10.13140/RG.2.2.23258.54728.

Campbell, Jeremy M. 2015. *Conjuring Property: Speculation and Environmental Futures in the Brazilian Amazon*. Culture, Place, and Nature: Studies in Anthropology and Environment. Seattle: University of Washington Press.

Carroll, Sean Goulding. 2021. “Airline Industry Charts Path to Carbon Neutrality by 2050.” *EURACTIV*, February 15. https://www.euractiv.com/section/aviation/news/airline-industry-charts-path-to-carbon-neutrality-by-2050/.

Case, Anne, and Angus Deaton. 2015. “Rising Morbidity and Mortality in Midlife among White Non-Hispanic Americans in the 21st Century.” *Proceedings of the National Academy of Sciences* 112 (49): 15078–83. https://doi.org/10.1073/pnas.1518393112.

———. 2020. *Deaths of Despair and the Future of Capitalism*. Princeton, NJ: Princeton University Press.

Center for Sustainable Energy. 2020. “Value Stacking with Distributed Energy Resources.” April 5. https://energycenter.org/thought-leadership/blog/value-stacking-distributed-energy-resources.

Central Statistics Office. 2018. "Milk Statistics." January 31. https://www.cso.ie/en/releasesandpublications/er/ms/milkstatisticsdecember2017/.

Chan, H. Ron, B. Andrew Chupp, Maureen L. Cropper, and Nicholas Z. Muller. 2018. "The Impact of Trading on the Costs and Benefits of the Acid Rain Program." *Journal of Environmental Economics and Management* 88 (March): 180–209. https://doi.org/10.1016/j.jeem.2017.11.004.

Chandler, Alfred D., Jr. 1993. *The Visible Hand: The Managerial Revolution in American Business.* Cambridge, MA: Harvard University Press. https://doi.org/10.2307/j.ctvjghwrj.

Chayes, Abram, and Antonia H. Chayes. 1991. "Adjustment and Compliance Processes in International Regulatory Regimes." In *Preserving the Global Environment: The Challenge of Shared Leadership*, edited by Jessica Tuchman Mathews, 280–308. New York: W. W. Norton.

———. 1998. *The New Sovereignty: Compliance with International Regulatory Agreements.* Cambridge, MA: Harvard University Press.

Cherry, Jeff (EPA engineer). 2016. Interview with Lauren Packard.

Chestnut, Lauraine G., and David M. Mills. 2005. "A Fresh Look at the Benefits and Costs of the US Acid Rain Program." *Journal of Environmental Management* 77 (3): 252–66. https://doi.org/10.1016/j.jenvman.2005.05.014.

China. 2015. "G20 Voluntary Peer Review by China and the United States on Inefficient Fossil Fuel Subsidies That Encourage Wasteful Consumption: China Self-Review Report." Paris: Organization for Economic Cooperation and Development. http://www.oecd.org/fossil-fuels/publication/G20%20China%20Self%20Review%20on%20Fossil%20Fuel%20Subsidies-China%20Self-report-20160902_English.pdf.

China, Germany, Mexico, and Organization for Economic Cooperation and Development. 2016. "The United States' Efforts to Phase Out and Rationalise Its Inefficient Fossil-Fuel Subsidies: A Report on the G20 Peer-Review of Inefficient Fossil-Fuel Subsidies That Encourage Wasteful Consumption in the United States." Paris: Organization for Economic Cooperation and Development. http://www.oecd.org/fossil-fuels/publication/United%20States%20Peer%20review_G20_FFS_Review_final_of_20160902.pdf.

Christensen, Clayton M. 2016. *The Innovator's Dilemma: When New Technologies Cause Great Firms to Fail.* Management of Innovation and Change Series. Boston: Harvard Business Review Press.

Clarke, Leon, Keigo Akimoto, Kejun Jiang, Mustafa Babiker, Geoffrey Blanford, Karen Fisher-Vanden, Jean-Charles Hourcade, et al. 2014. "Assessing Transformation Pathways." In *Climate Change 2014: Mitigation of Climate Change. Contribution of Working Group III to the Fifth Assessment Report of the Intergovernmental Panel on Climate Change*, edited by Ottmar Edenhofer, Ramón Pichs-Madruga, Youba Sokona, Ellie Farahani, Susanne Kadner, Kristin Seyboth, Anna Adler, et al., 413–510. Cambridge: Cambridge University Press.

Clean Air Act. 1963. 42 U.S.C. *§7413(a)(5).*

Cohen, Armond, Arne Olson, Clea Kolster, David G. Victor, Ejeong Baik, Jane C. S. Long, Jesse D. Jenkins, et al. 2021. "Clean Firm Power Is the Key to California's Carbon-Free Energy Future." *Issues in Science and Technology*, March. https://issues.org/california-decarbonizing-power-wind-solar-nuclear-gas/.

Cohen, Linda R., and Roger G. Noll, eds. 1991. *The Technology Pork Barrel.* Washington, DC: Brookings Institution Press.

Cohen, Wesley M., and Daniel A. Levinthal. 1990. "Absorptive Capacity: A New Perspective on Learning and Innovation." *Administrative Science Quarterly*, 128–52.

Coons, Chris, and Scott Peters. 2021. "S.2378—FAIR Transition and Competition Act." https://www.congress.gov/bill/117th-congress/senate-bill/2378/titles.

Copeland, Claudia. 2012. "Clean Water Act and Pollutant Total Maximum Daily Loads (TMDLs)." Report for Congress, September 21, R42752. Washington, DC: Congressional Research Service.

Correa, Paulo Guilherme, and Cristiane Alkmin Junqueira Schmidt. 2014. "Public Research Organizations and Agricultural Development in Brazil: How Did Embrapa Get It Right?" No. 88490. World Bank. https://documents.worldbank.org/pt/publication/documents-reports/documentdetail/156191468236982040/public-research-organizations-and-agricultural-development-in-brazil-how-did-embrapa-get-it-right.

Cosens, Barbara A., J. B. Ruhl, Niko Soininen, and Lance Gunderson. 2020. "Designing Law to Enable Adaptive Governance of Modern Wicked Problems." *Vanderbilt Law Review* 73 (6): 1687–732.

Coslovsky, Salo V. 2014. "Economic Development without Pre-Requisites: How Bolivian Producers Met Strict Food Safety Standards and Dominated the Global Brazil-Nut Market." *World Development* 54 (February): 32–45. https://doi.org/10.1016/j.worlddev.2013.07.012.

———. 2021. "Oportunidades para Exportação de Produtos Compatíveis Com a Floresta na Amazônia Brasileira." Amazônia 2030. https://amazonia2030.org.br/oportunidades-para-exportacao-de-produtos-compativeis-com-a-floresta-na-amazonia-brasileira/.

Cullenward, Danny, and David G. Victor. 2020. *Making Climate Policy Work*. Cambridge, UK: Polity Press.

Dairy Industry Prospectus Report. 2003. "Strategic Development Plan for the Irish Dairy Processing Sector." https://www.agriculture.gov.ie/publications/2000-2003/dairyindustryprospectusreport2003/.

Daly, Donal, Marie Archbold, and Jenny Deakin. 2016. "Progress and Challenges in Managing Our Catchments Effectively." *Biology and Environment: Proceedings of the Royal Irish Academy* 116b (3): 157–66.

de Búrca, Gráinne, and Joanne Scott. 2006. *Law and New Governance in the EU and the US*. Cumnor, UK: Hart Publishing. https://doi.org/10.5040/9781472563668.

de Paiva Serôa da Motta, Raquel, Cintia Munch Cavalcanti, Joyce Brandão, Mariana Pereira, Paulo Lima, and Beatriz Domeniconi. 2018. "Small Scale, Great Opportunity: Towards Sustainable Young Livestock Farming in the Amazon and the Potential of the Innovation and Learning Hubs (ILHs): A Study to Analyse the Potential of Implantation of Innovation and Learning Hubs (ILHs) for the Dissemination of Innovative Practices in Livestock Farming for the Mitigation of Greenhouse Gases." Wageningen, Netherlands: Wageningen Livestock Research. https://doi.org/10.18174/495347.

DeSombre, Elizabeth R., and Joanne Kauffman. 1996. "The Montreal Protocol Multilateral Fund: Partial Success Story." In *Institutions for Environmental Aid: Pitfalls and Promise*, edited by Robert O. Keohane and Marc A. Levy, 89–126. Global Environmental Accords. Cambridge, MA: MIT Press.

Dewey, John. 1927. *The Public and Its Problems*. New York: H. Holt and Company.

Dharmawan, Arya Hadi, Dyah Ita Mardiyaningsih, Faris Rahmadian, Bayu Eka Yulian, Heru Komarudin, Pablo Pacheco, Jaboury Ghazoul, and Rizka Amalia. 2021. "The Agrarian, Structural and Cultural Constraints of Smallholders' Readiness for Sustainability Standards Implementation: The Case of Indonesian Sustainable Palm Oil in East Kalimantan." *Sustainability* 13 (5): 2611. https://doi.org/10.3390/su13052611.

Dixon, Lloyd, Steven Garber, and Mary E. Vaiana. 1996. *Making ZEV Policy despite Uncertainty: An Annotated Briefing for the California Air Resources Board*. Santa Monica, CA: Rand Corporation.

Doblinger, Claudia, Kavita Surana, and Laura Diaz Anadon. 2019. "Governments as Partners: The Role of Alliances in U.S. Cleantech Startup Innovation." *Research Policy* 48 (6): 1458–75. https://doi.org/10.1016/j.respol.2019.02.006.

Dolphin, Geoffroy, Michael G. Pollitt, and David M. Newbery. 2020. "The Political Economy of Carbon Pricing: A Panel Analysis." *Oxford Economic Papers* 72, no. 2 (April): 472–500. https://doi.org/10.1093/oep/gpz042.

Dorf, Michael C., and Charles F. Sabel. 1998. "A Constitution of Democratic Experimentalism." *Columbia Law Review* 98 (2): 267–473.

Downs, George W., David M. Rocke, and Peter N. Barsoom. 1996. "Is the Good News about Compliance Good News about Cooperation?" *International Organization* 50 (3): 379–406.

Drezner, Daniel W. 2009. "The Power and Peril of International Regime Complexity." *Perspectives on Politics* 7 (1): 65–70. https://doi.org/10.1017/S1537592709090100.

Electric Power Research Institute (EPRI). 2005. "EPRI Honors Southern Company and Georgia Power's Plant Yates for Leadership in Mercury Emissions Control Technology." http://mydocs.epri.com/docs/CorporateDocuments/Newsroom/060205_YatesPlant.html.

Ellerman, Denny. 2010. "US Experience with Emissions Trading: Lessons for CO_2 Emissions Trading." In *Emissions Trading for Climate Policy: US and European Perspectives*, edited by Bernd Hansjürgens, 78–95. Cambridge: Cambridge University Press.

Elliott, E. Donald. 1992. "Re-Inventing Rulemaking." *Duke Law Journal* 41:1490–96.

Energy Transitions Commission. 2018. "Mission Possible: Reaching Net-Zero Carbon Emissions from Harder-to-Abate Sectors by Mid-Century." London: Energy Transitions Commission. http://www.energy-transitions.org/sites/default/files/ETC_MissionPossible_FullReport.pdf.

———. 2020a. "Making Mission Possible: Delivering a Net-Zero Economy." London: Energy Transitions Commission. https://www.energy-transitions.org/publications/making-mission-possible/.

———. 2020b. "Towards a Low-Carbon Steel Sector: Overview of the Changing Market, Technology and Policy Context for Indian Steel." London: Energy Transitions Commission. https://www.energy-transitions.org/publications/towards-a-low-carbon-steel-sector/.

Environment and Social Development Organization (ESDO). 2018. "Movement for Amalgam Phase Down Turned Strong in Global Workshop." May 14. https://esdo.org/movement-for-amalgam-phase-down-turned-strong-in-global-workshop/.

Environmental Defense Fund. 2018. "How Economics Solved Acid Rain." September. https://www.edf.org/approach/markets/acid-rain.

Environmental Protection Agency (EPA). 1990. "1990 Clean Air Act Amendment Summary." https://www.epa.gov/clean-air-act-overview/1990-clean-air-act-amendment-summary.

EPA Catchments Unit. 2018. "The Local Authority Waters Programme Catchment Assessment Team." Catchments.ie, December 11. https://www.catchments.ie/the-local-authority-waters-programme-catchment-assessment-team/.

European Commission. 2010. "Commission Staff Working Document: Accompanying Document to the Commission Decision on Applying Use Restrictions on International Credits (from HFC-23 and N2O Projects) Pursuant to Article 11a(9) of Directive 2009/29/EC." Brussels: SEC. https://ec.europa.eu/clima/sites/clima/files/ets/markets/docs/sec_2010_yyy_en_0.pdf.

European Union (EU). 2013. "Commission Regulation (EU) No. 1123/2013 of 8 November 2013 on Determining International Credit Entitlements Pursuant to Directive 2003/87/EC of the European Parliament and of the Council." *Official Journal of European Union*. https://eur-lex.europa.eu/LexUriServ/LexUriServ.do?uri=OJ:L:2013:299:0032:0033:EN:PDF.

Eurostat. 2017. "Milk and Milk Product Statistics." https://ec.europa.eu/eurostat/statistics-explained/index.php/Milk_and_milk_product_statistics.

Falkner, Robert. 2016. "The Paris Agreement and the New Logic of International Climate Politics." *International Affairs* 92 (5): 1107–25. https://doi.org/10.1111/1468-2346.12708.

Fama, Eugene F., and Michael C. Jensen. 1983. "Agency Problems and Residual Claims." *Journal of Law and Economics* 26 (2): 327–49.

Fishkin, James S. 2018. *Democracy When the People Are Thinking: Revitalizing Our Politics through Public Deliberation*. Oxford: Oxford University Press.

Fishkin, James S., Alice Siu, Larry Diamond, and Norman Bradburn. 2021. "Is Deliberation an Antidote to Extreme Partisan Polarization? Reflections on 'America in One Room.'" *American Political Science Review*, 1–18. https://doi.org/10.1017/S0003055421000642.

Fishman, Akiva, Edegar Oliveira, and Lloyd Gamble. 2017. "Tackling Deforestation through a Jurisdictional Approach: Lessons from the Field." World Wildlife Fund. https://wwfint.awsassets.panda.org/downloads/wwf_jurisdictional_approaches_fullpaper_web_1.pdf.

Fitzgerald, Ciaran. 2019. "Dairy in the Irish Economy!" *Irish Dairying: Growing Sustainably*, 46. Cork, Ireland: Moorepark Food Research Centre.

Fitzgerald, Garrett, James Mandel, Jesse Morris, and Hervé Touati. 2015. "The Economics of Battery Energy Storage: How Multi-Use, Customer-Sited Batteries Deliver the Most Services and Value to Customers and the Grid." Rocky Mountain Institute. https://rmi.org/2015_10_07_year_of_the_battery_but_storage_can_do_much_more/.

Food and Agriculture Organization. 2018. "Dairy Market Review: Milk Production Trends in 2017." Rome: Food and Agriculture Organization of the United Nations.

Freitas, Josimar da Silva, Marcos Rodrigues, and David Costa Correia Silva. 2016. "Áreas Protegidas en el Amazon: Un Análisis Institucional Extractiva Reserva el Alto Jurua." *Contribuciones a Las Ciencias Sociales* (3).

Gardner, Sarah. 2014. "LA Smog: The Battle against Air Pollution." *Marketplace*, July. http://www.marketplace.org/2014/07/14/sustainability/we-used-be-china/la-smog-battle-against-air-pollution.

Gecko Project. 2017. "The Making of a Palm Oil Fiefdom." October 11. https://thegeckoproject.org/the-making-of-a-palm-oil-fiefdom-7e1014e8c342.

Geman, Ben. 2021. "John Kerry Says EU Carbon Border Tax Adjustment Should Be a 'Last Resort.'" *Axios*, March 12. https://www.axios.com/kerry-eu-carbon-border-tax-adjustment-94bfdf7a-d74c-4d78-8035-1bb680800aea.html.

Gereffi, Gary, Marino Regini, and Charles F. Sabel. 2014. "On Richard M. Locke, *The Promise and Limits of Private Power: Promoting Labor Standards in a Global Economy*, New York, Cambridge University Press, 2013." *Socio-Economic Review* 12 (1): 219–35. https://doi.org/10.1093/ser/mwt023.

Germany, Indonesia, United States, International Monetary Fund, and Organization for Economic Cooperation and Development. 2016. "China's Efforts to Phase Out and Rationalise Its Inefficient Fossil-Fuel Subsidies: A Report on the G20 Peer Review of Inefficient Fossil-Fuel Subsidies That Encourage Wasteful Consumption in China." Paris: Organization for Economic Cooperation and Development. http://www.oecd.org/fossil-fuels/publication/G20%20China%20Peer%20Review_G20_FFS_Review_final_of_20160902.pdf.

Getachew, Adom. 2019. *Worldmaking after Empire: The Rise and Fall of Self-Determination*. Princeton, NJ: Princeton University Press.

Giakoumis, Theodoros, and Nikolaos Voulvoulis. 2018. "Progress with Monitoring and Assessment in the WFD Implementation in Five European River Basins: Significant Differences but Similar Problems." *European Journal of Environmental Sciences* 8 (1): 44–50. https://doi.org/10.14712/23361964.2018.7.

Gibbs, Holly K., Jacob Munger, Jessica L'Roe, Paulo Barreto, Ritaumaria Pereira, Matthew Christie, Ticiana Amaral, and Nathalie F. Walker. 2016. "Did Ranchers and Slaughterhouses Respond to Zero-Deforestation Agreements in the Brazilian Amazon?: Brazil's Zero-Deforestation Pacts." *Conservation Letters* 9 (1): 32–42. https://doi.org/10.1111/conl.12175.

Gilson, Ronald J., and Jeffrey N. Gordon. 2020. "Board 3.0: What the Private-Equity Governance Model Can Offer Public Companies." *Journal of Applied Corporate Finance* 32 (3): 43–51. https://doi.org/10.1111/jacf.12417.

Gilson, Ronald J., Charles F. Sabel, and Robert E. Scott. 2009. "Contracting for Innovation: Vertical Disintegration and Interfirm Collaboration." *Columbia Law Review* 109 (3): 431–502.

———. 2014. "Text and Context: Contract Interpretation as Contract Design." *Cornell Law Review* 100:23–97.

Gomes, Carlos Valério Aguiar, Stephen G. Perz, and Jacqueline Michelle Vadjunec. 2012. "Convergence and Contrasts in the Adoption of Cattle Ranching: Comparisons of Smallholder Agriculturalists and Forest Extractivists in the Amazon." *Journal of Latin American Geography* 11 (1): 99–120. https://doi.org/10.1353/lag.2012.0018.

Goodin, Robert E., and Simon J. Niemeyer. 2003. "When Does Deliberation Begin? Internal Reflection versus Public Discussion in Deliberative Democracy." *Political Studies* 51 (4): 627–49. https://doi.org/10.1111/j.0032-3217.2003.00450.x.

Governor's Office of Planning and Research. 2012. "California Solar Permitting Guidebook: Improving Permit Review and Approval for Small Solar Photovoltaic (PV) Systems." Solar Permitting Work Group. https://www.buildingincalifornia.com/wp-content/uploads/2014/02/CaliforniaSolarPermittingGuidebook-2012.pdf.

———. 2014. "California Solar Permitting Guidebook: Improving Permit Review and Approval for Small Solar Systems." Solar Permitting Task Force. https://energycenter.org/sites/default/files/docs/nav/policy/research-and-reports/California_Solar_Permitting_Guidebook_2014.pdf.

Greene, Owen. 1998. "The System for Implementation Review in the Ozone Regime." In *The Implementation and Effectiveness of International Environmental Commitments: Theory and Practice*, edited by David G. Victor, Kal Raustiala, and Eugene B. Skolnikoff, 89–136. Cambridge, MA: MIT Press.

G20 Working Group on Energy and Commodities. 2012. "G20 Inefficient Fossil Subsidy Reform Peer Review Process." Los Cabos, Mexico: G20.

Guanziroli, Carlos, Antônio Márcio Buainain, Gabriela Benatti, and Vahíd Shaikhzadeh Vahdat. 2019. "The Fate of Family Farming under the New Pattern of Agrarian Development in Brazil." In *Agricultural Development in Brazil: The Rise of a Global Agro-Food Power*, edited by Antônio Márcio Buainain, Rodrigo Lanna, and Zander Navarro, 174–88. New York: Routledge. https://doi.org/10.4324/9781351029742.

Gunningham, Neil, Robert A. Kagan, and Dorothy Thornton. 2004. "Social License and Environmental Protection: Why Businesses Go beyond Compliance." *Law and Social Inquiry* 29 (2): 307–41. https://doi.org/10.1111/j.1747-4469.2004.tb00338.x.

Haas, Peter M., Marc A. Levy, and Edward A. Parson. 1992. "How Should We Judge UNCED's Success?" *Environment: Science and Policy for Sustainable Development* 34 (8): 6–33. https://doi.org/10.1080/00139157.1992.9931467.

Hahn, Robert W., and Gordon L. Hester. 1989. "Where Did All the Markets Go? An Analysis of EPA's Emissions Trading Program." *Yale Journal on Regulation* 6 (1): 109–53.

Hale, Thomas. 2016. "'All Hands on Deck': The Paris Agreement and Nonstate Climate Action." *Global Environmental Politics* 16 (3): 12–22. https://doi.org/10.1162/GLEP_a_00362.

———. 2017. "Climate Change: From Gridlock to Catalyst." In *Beyond Gridlock*, edited by Thomas Hale and David Held. Cambridge, UK: Polity Press.

———. 2020. "Catalytic Cooperation." *Global Environmental Politics* 20 (4): 73–98. https://doi.org/10.1162/glep_a_00561.

Hale, Thomas, Sander Chan, Angel Hsu, Andrew Clapper, Cynthia Elliott, Pedro Faria, Takeshi Kuramochi, et al. 2021. "Sub- and Non-State Climate Action: A Framework to Assess Progress, Implementation and Impact." *Climate Policy* 21 (3): 406–20. https://doi.org/10.1080/14693062.2020.1828796.

Hale, Thomas, David Held, and Kevin Young. 2017. *Gridlock: Why Global Cooperation Is Failing When We Need It Most*. Cambridge, UK: Polity Press.

Hanna, Ryan, Vahid R. Disfani, Hamed Valizadeh Haghi, David G. Victor, and Jan Kleissl. 2019. "Improving Estimates for Reliability and Cost in Microgrid Investment Planning Models." *Journal of Renewable and Sustainable Energy* 11 (4): 045302. https://doi.org/10.1063/1.5094426.

Hart, David, and Alfred Sarkissian. 2016. "Deployment of Grid-Scale Batteries in the United States." Prepared for Office of Energy Policy and Systems Analysis, US Department of Energy. https://www.energy.gov/sites/prod/files/2017/01/f34/Deployment%20of%20Grid-Scale%20Batteries%20in%20the%20United%20States.pdf.

Hathaway, Oona A., Curtis Bradley, and Jack Landman Goldsmith. 2020. "The Failed Transparency Regime for Executive Agreements: An Empirical and Normative Analysis." *Harvard Law Review* 134 (2): 629–725.

Hawkins, Darren G., David A. Lake, Daniel L. Nielson, and Michael J. Tierney, eds. 2006. *Delegation and Agency in International Organizations*. Political Economy of Institutions and Decisions. Cambridge: Cambridge University Press.

Hayek, Friedrich A. 1949. *Individualism and Economic Order*. London: Routledge and Kegan Paul.

Hecht, Alan D., and Dennis Tirpak. 1995. "Framework Agreement on Climate Change: A Scientific and Policy History." *Climatic Change* 29 (4): 371–402. https://doi.org/10.1007/BF01092424.

Hecker, JayEtta Z. 2007. "Freight Railroads: Updated Information on Rates and Competition Issues." GAO-07-1245T, September 25. Washington, DC: US Government Accountability Office.

Heilmann, Sebastian. 2008. "From Local Experiments to National Policy: The Origins of China's Distinctive Policy Process." *China Journal* 59 (January): 1–30. https://doi.org/10.1086/tcj.59.20066378.

Heilmayr, Robert. 2019. "Replication Data for Brazil's Amazon Soy Moratorium Reduced Deforestation." Harvard Dataverse. https://doi.org/10.7910/DVN/LE42B1.

Henderson, Rebecca. 2020. *Reimagining Capitalism in a World on Fire*. New York: PublicAffairs.

Hering, Daniel, Angel Borja, Jacob Carstensen, Laurence Carvalho, Mike Elliott, Christian K. Feld, Anna-Stiina Heiskanen, Richard K. Johnson, Jannicke Moe, and Didier Pont. 2010. "The European Water Framework Directive at the Age of 10: A Critical Review of the Achievements with Recommendations for the Future." *Science of the Total Environment* 408 (19): 4007–19. https://doi.org/10.1016/j.scitotenv.2010.05.031.

Hirschman, Albert O. 2014. *Development Projects Observed*. Washington, DC: Brookings Institution Press. http://site.ebrary.com/id/10994719.

Hiscox, Michael J., and Nicholas F. B. Smyth. 2006. "Is There Consumer Demand for Improved Labor Standards? Evidence from Field Experiments in Social Labeling." Department of Government, Harvard University.

Ho, Daniel E. 2017. "Does Peer Review Work? An Experiment of Experimentalism." *Stanford Law Review* 69 (February): 1–119.

Hochstetler, Kathryn, and Margaret E. Keck. 2007. *Greening Brazil: Environmental Activism in State and Society*. Durham, NC: Duke University Press.

Hockenos, Paul. 2015. "The History of the Energiewende." Clean Energy Wire, June 22. https://www.cleanenergywire.org/dossiers/history-energiewende.

Hoekman, Bernard. 2016. "The Bali Trade Facilitation Agreement and Rulemaking in the WTO: Milestone, Mistake, or Mirage?" In *The World Trade System: Trends and Challenges*, edited by Jagdish N. Bhagwati, Pravin Krishna, and Arvind Panagariya, 149–92. Cambridge, MA: MIT Press.

Hoekman, Bernard, and Charles F. Sabel. 2019. "Open Plurilateral Agreements, International Regulatory Cooperation and the WTO." *Global Policy* 10 (3): 297–312. https://doi.org/10.1111/1758-5899.12694.

Hoelle, Jeffrey. 2014. "Cattle Culture in the Brazilian Amazon." *Human Organization* 73 (4): 363–74. https://doi.org/10.17730/humo.73.4.u61u675428341165.

———. 2015. *Rainforest Cowboys: The Rise of Ranching and Cattle Culture in Western Amazonia.* Austin: University of Texas Press.

Hoffmann, Matthew J. 2005. *Ozone Depletion and Climate Change: Constructing a Global Response.* SUNY Series in Global Politics. Albany: SUNY Press.

———. 2011. *Climate Governance at the Crossroads: Experimenting with a Global Response after Kyoto.* Oxford: Oxford University Press.

Hofmann, David J., Terry L. Deshler, Patrick Aimedieu, William A. Matthews, Peter V. Johnston, Yutaka Kondo, Wendy R. Sheldon, Gregory J. Byrne, and James R. Benbrook. 1989. "Stratospheric Clouds and Ozone Depletion in the Arctic during January 1989." *Nature* 340 (6229): 117–21. https://doi.org/10.1038/340117a0.

Hofmeister, Naira. 2020. "Conflict in the Chico Mendes Reserve Threatens This Pioneering Amazonian Project." Translated by Matt Rinaldi. Mongabay, January 16. https://news.mongabay.com/2020/01/conflict-in-the-chico-mendes-reserve-threaten-this-pioneering-amazonian-project/.

Hoikkala, Hanna, and Jesper Starn. 2020. "Sweden Moves Closer to Making Fossil-Fuel-Free Steel." *Bloomberg Green*, August 31. https://www.bloomberg.com/news/articles/2020-08-31/sweden-advances-on-road-to-fossil-free-steel-with-three-way-jv.

Holling, C. S., ed. 1978. *Adaptive Environmental Assessment and Management.* Chichester, UK: John Wiley.

Holman, Jacqueline. 2020. "Global Passenger EV Sales Forecast to Reach 6.2 Million Units by 2024: MI." S&P Global, July 31. https://www.spglobal.com/platts/en/market-insights/latest-news/metals/073120-global-passenger-ev-sales-forecast-to-reach-62-million-units-by-2024-mi.

Holmstrom, Bengt. 1982. "Moral Hazard in Teams." *Bell Journal of Economics* 13 (2): 324–40. https://doi.org/10.2307/3003457.

Holmstrom, Bengt, and Paul Milgrom. 1991. "Multitask Principal-Agent Analyses: Incentive Contracts, Asset Ownership, and Job Design." *Journal of Law, Economics, and Organization* 7:24–52.

Horn, Miriam. 2017. *Rancher, Farmer, Fisherman: Conservation Heroes of the American Heartland.* New York: W. W. Norton.

Howell, Sabrina T. 2017. "Financing Innovation: Evidence from R&D Grants." *American Economic Review* 107 (4): 1136–64. https://doi.org/10.1257/aer.20150808.

Hughes, Paul (CARB emissions compliance manager). 2015. Interview with Lauren Packard.

Hutchins, Edwin. 2000. *Cognition in the Wild.* Cambridge, MA: MIT Press.

"IMO Agrees to CO2 Emissions Target." 2018. Maritime Executive, April 13. https://www.maritime-executive.com/article/imo-agrees-to-co2-emissions-target.

Intergovernmental Panel on Climate Change (IPCC). 1990. *Climate Change: IPCC First Assessment Report, Working Group I.* Edited by Intergovernmental Panel on Climate Change, John Theodore Houghton, G. J. Jenkins, and J. J. Ephraums. Cambridge: Cambridge University Press.

International Energy Agency. 2012. "Gas Pricing: China's Challenges and IEA Experience." Partner Country Series. Paris: International Energy Agency. https://www.iea.org/reports/partner-country-series-gas-pricing-chinas-challenges-and-iea-experience.

———. 2020a. *Energy Technology Perspectives 2020—Special Report on Clean Energy Innovation: Accelerating Technology Progress for a Sustainable Future.* Paris: International Energy Agency. https://www.iea.org/reports/energy-technology-perspectives-2020.

———. 2020b. "World Energy Outlook 2020." Paris: International Energy Agency. https://www.iea.org/reports/world-energy-outlook-2020.

International Harvester Co. v. Ruckelshaus. 1973. 478 F.2d 615 (D.C. Cir.).

"The Invisible Green Hand." 2002. *Economist*, July 6. https://www.economist.com/special-report/2002/07/06/the-invisible-green-hand.

Ivanova, Maria. 2021. *The Untold Story of the World's Leading Environmental Institution: UNEP at Fifty*. Cambridge, MA: MIT Press. https://doi.org/10.7551/mitpress/12373.001.0001.

Jackson, Rachel. 2015. "A Credible Commitment: Reducing Deforestation in the Brazilian Amazon, 2003–2012." Princeton University, Innovations for Successful Societies. https://successfulsocieties.princeton.edu/publications/credible-commitment-reducing-deforestation-brazilian-amazon-2003%E2%80%932012.

Jacobson, Dan. 2020. "Three Lessons from California's Million Solar Roofs Milestone." Environment California. January 7. https://environmentcalifornia.org/blogs/blog/cae/three-lessons-californias-million-solar-roofs-milestone.

Jacobsson, Staffan, and Volkmar Lauber. 2006. "The Politics and Policy of Energy System Transformation—Explaining the German Diffusion of Renewable Energy Technology." *Energy Policy* 34 (3): 256–76. https://doi.org/10.1016/j.enpol.2004.08.029.

Jaramillo-Giraldo, Carolina, Britaldo Soares Filho, Sónia M. Carvalho Ribeiro, and Rivadalve Coelho Gonçalves. 2017. "Is It Possible to Make Rubber Extraction Ecologically and Economically Viable in the Amazon? The Southern Acre and Chico Mendes Reserve Case Study." *Ecological Economics* 134 (April): 186–97. https://doi.org/10.1016/j.ecolecon.2016.12.035.

Jennison, Michael. 2013. "The Future of Aviation Safety Regulation: New US-EU Agreement Harmonizes and Consolidates the Transatlantic Regime, but What Is the Potential for Genuine Regulatory Reform?" *Air and Space Law* 38 (4–5): 333–50.

Kahler, Miles. 1992. "Multilateralism with Small and Large Numbers." *International Organization* 46 (3): 681–708.

Kane, Mark. 2020. "California: Plug-in Electric Car Sales Down 17% in H1 2020." *InsideEVs*, August 12. https://insideevs.com/news/438647/california-plugin-car-sales-h1-2020/.

Keohane, Robert O., and David G. Victor. 2011. "The Regime Complex for Climate Change." *Perspectives on Politics* 9 (1): 7–23.

Kerr, William A., and Jill E. Hobbs. 2021. "18. Is NAFTA's Northern Border Thickening for Agri-Food Products?" In *Navigating a Changing World*, edited by Geoffrey Hale and Greg Anderson, 430–52. Toronto: University of Toronto Press. https://doi.org/10.3138/9781487537708-020.

Kessler, Jeremy, and Charles F. Sabel. 2021. "The Uncertain Future of Administrative Law." *Daedalus* 150 (3): 188–207. https://doi.org/10.1162/daed_a_01867.

Kinghorn, Jonathan. 2016. "Floods in Thailand Are Regular Natural Disasters." *AIR* (blog). December 12. https://www.air-worldwide.com/Blog/Floods-in-Thailand-Are-Regular-Natural-Disasters/.

Knight, Frank H. 1921. *Risk, Uncertainty, and Profit*. Vol. 31. Boston: Houghton Mifflin Co.

Krastev, Ivan, and Stephen Holmes. 2020. *The Light That Failed: A Reckoning*. London: Penguin Books.

Kröger, Markus. 2020. "Deforestation, Cattle Capitalism and Neodevelopmentalism in the Chico Mendes Extractive Reserve, Brazil." *Journal of Peasant Studies* 47 (3): 464–82. https://doi.org/10.1080/03066150.2019.1604510.

Lang, Winfried. 1991. "Is the Ozone Depletion Regime a Model for an Emerging Regime on Global Warming?" *UCLA Journal of Environmental Law and Policy* 9 (2): 161–74.

Larson, Eric, Chris Greig, Jesse Jenkins, Erin Mayfield, Andrew Pascale, Chuan Zhang, Joshua Drossman, Robert Williams, Steve Pacala, and Robert Socolow. 2020. "Net-Zero America: Potential Pathways, Infrastructure, and Impacts." Interim report, Princeton University, Princeton, NJ, December 15. https://environmenthalfcentury.princeton.edu/sites/g/files/toruqf331/files/2020-12/Princeton_NZA_Interim_Report_15_Dec_2020_FINAL.pdf.

Latour, Bruno. 1999. *Pandora's Hope: Essays on the Reality of Science Studies*. Cambridge, MA: Harvard University Press.

Lauber, Volkmar, and Lutz Mez. 2004. "Three Decades of Renewable Electricity Policies in Germany." *Energy and Environment* 15 (4): 599–623. https://doi.org/10.1260/0958305042259792.

Lavenex, Sandra. 2015. "Experimentalist Governance in EU Neighbourhood Policies." In *Extending Experimentalist Governance? The European Union and Transnational Regulation*, edited by Jonathan Zeitlin, 22–47. Oxford: Oxford University Press. https://doi.org/10.1093/acprof:oso/9780198724506.003.0002.

Lee, Kai N. 1993. *Compass and Gyroscope: Integrating Science and Politics for the Environment.* Washington, DC: Island Press.

Lester, Richard K., and David M. Hart. 2012. *Unlocking Energy Innovation: How America Can Build a Low-Cost, Low-Carbon Energy System.* Cambridge, MA: MIT Press.

Lindsay, Steven J. 2018. "Timing Judicial Review of Agency Interpretations in Chevron's Shadow." *Yale Law Journal* 127 (8): 2204–585.

Lipscomb, Molly, and Niveditha Prabakaran. 2020. "Property Rights and Deforestation: Evidence from the Terra Legal Land Reform in the Brazilian Amazon." *World Development* 129 (89): 104854. https://doi.org/10.1016/j.worlddev.2019.104854.

Locke, Richard M. 2013. *The Promise and Limits of Private Power: Promoting Labor Standards in a Global Economy.* Cambridge: Cambridge University Press. https://doi.org/10.1017/CBO9781139381840.

Lucas, Anton E., and Carol Warren. 2013. *Land for the People: The State and Agrarian Conflict in Indonesia.* Ohio University Research in International Studies. Southeast Asia Series, no. 126. Athens: Ohio University Press.

Luskin, Robert C., Ian O'Flynn, James S. Fishkin, and David Russell. 2014. "Deliberating across Deep Divides." *Political Studies* 62 (1): 116–35. https://doi.org/10.1111/j.1467-9248.2012.01005.x.

Luskin, Robert C., James S. Fishkin, and Roger Jowell. 2002. "Considered Opinions: Deliberative Polling in Britain." *British Journal of Political Science* 32 (3): 455–87. https://doi.org/10.1017/S0007123402000194.

MacKinnon, Duncan. 2017. "A Battery for SA: Durable or Barely Energised?" Australian Energy Council, July 14. https://www.energycouncil.com.au/analysis/a-battery-for-sa-durable-or-barely-energised/.

Magnusson, Thomas, and Christian Berggren. 2011. "Entering an Era of Ferment—Radical vs Incrementalist Strategies in Automotive Power Train Development." *Technology Analysis and Strategic Management* 23 (3): 313–30. https://doi.org/10.1080/09537325.2011.550398.

Mallet, Victor, and Roula Khalaf. 2020. "FT Interview: Emmanuel Macron Says It Is Time to Think the Unthinkable." *Financial Times*, April 16. https://www.ft.com/content/3ea8d790-7fd1-11ea-8fdb-7ec06edeef84.

"Managing Oversupply" 2021. California Independent System Operator, September 13. http://www.caiso.com/informed/Pages/ManagingOversupply.aspx.

Martin, Paul. 2020. "The G20 Today: Pandemic Disease, Climate Change, and the Need for a Rules-Based Order." Calgary: Canadian Global Affairs Institute. https://d3n8a8pro7vhmx.cloudfront.net/cdfai/pages/4402/attachments/original/1586037423/The_G20_Today_Pandemic_Disease__Climate_Change__and_the_Need_for_a_Rules-Based_Order.pdf?1586037423.

Mazzucato, Mariana. 2013. *The Entrepreneurial State: Debunking Public vs. Private Sector Myths.* Anthem Frontiers of Global Political Economy. London: Anthem Press.

McCarthy, Michael (CARB chief technology officer, mobile source emissions). 2015. Interview with Lauren Packard.

Meckling, Jonas, and Eric Biber. 2021. "A Policy Roadmap for Negative Emissions Using Direct Air Capture." *Nature Communications* 12 (1): 2051. https://doi.org/10.1038/s41467-021-22347-1.

Methyl Bromide Alternatives Outreach (MBAO). n.d. "What Is MBAO?" https://mbao.org.

Mildenberger, Matto. 2020. *Carbon Captured: How Business and Labor Control Climate Politics*. American and Comparative Environmental Policy. Cambridge, MA: MIT Press.

Mills, Russell W., and Dorit Rubinstein Reiss. 2014. "Secondary Learning and the Unintended Benefits of Collaborative Mechanisms: The Federal Aviation Administration's Voluntary Disclosure Programs: The Benefit of Secondary Learning." *Regulation and Governance* 8 (4): 437–54. https://doi.org/10.1111/rego.12046.

Morales, Alex. 2015. "Eight Lessons from the Climate Disaster and Why Paris Worked." *Bloomberg*, December 13. https://www.bloomberg.com/news/articles/2015-12-13/eight-lessons-from-the-climate-disaster-and-why-paris-worked.

Najam, Adil, Saleemul Huq, and Youba Sokona. 2003. "Climate Negotiations beyond Kyoto: Developing Countries Concerns and Interests." *Climate Policy* 3 (3): 221–31. https://doi.org/10.1016/S1469-3062(03)00057-3.

National Academies of Sciences, Engineering, and Medicine, 2017. *An Assessment of ARPA-E*. Washington, DC: The National Academies Press. https://doi.org/10.17226/24778.

———. 2021a. *Accelerating Decarbonization of the U.S. Energy System*. Washington, DC: The National Academies Press. https://doi.org/10.17226/25932.

———. 2021b. *The Future of Electric Power in the United States*. Washington, DC: The National Academies Press. https://doi.org/10.17226/25968.

National Research Council. 1983. "Risk Assessment in the Federal Government: Managing the Process." Washington, DC: National Research Council. https://doi.org/10.17226/366.

Natural Resources Defense Council, Inc. v. US Environmental Protection Agency. 1981. 655 F.2d 318, 329 (D.C. Cir.).

Nature Conservancy. 2018. "São Félix do Xingu, Brazil: A Jurisdictional Approach to Conserving the Amazon." Arlington, VA: Nature Conservancy. https://www.nature.org/content/dam/tnc/nature/en/documents/TNC_JurisdictionalApproaches_CaseStudies_Brazil.pdf.

Navarro, Zander. 2020. "A Economia Agropecuária do Brasil: A Grande Transformação." São Paulo: Baraúna.

Neblo, Michael A., Kevin M. Esterling, and David M. J. Lazer. 2019. "Democracy When the People Are Thinking: Revitalizing Our Politics through Public Deliberation. By James S. Fishkin. New York: Oxford University Press, 2018. 272p. $24.95 Cloth." *Perspectives on Politics* 17 (2): 529–31. https://doi.org/10.1017/S1537592719001312.

Nemet, Gregory F. 2019. *How Solar Energy Became Cheap: A Model for Low-Carbon Innovation*. London: Routledge.

Nepstad, Daniel, David McGrath, Claudia Stickler, Ane Alencar, Andrea Azevedo, Briana Swette, Tathiana Bezerra, et al. 2014. "Slowing Amazon Deforestation through Public Policy and Interventions in Beef and Soy Supply Chains." *Science* 344 (6188): 1118–23. https://doi.org/10.1126/science.1248525.

Newell, Richard G., and Kristian Rogers. 2003. "The U.S. Experience with the Phasedown of Lead in Gasoline." Washington, DC: Resources for the Future.

Nordhaus, William D. 2015. "Climate Clubs: Overcoming Free-Riding in International Climate Policy." *American Economic Review* 105 (4): 1339–70. https://doi.org/10.1257/aer.15000001.

———. 2021. *The Spirit of Green: The Economics of Collisions and Contagions in a Crowded World*. Princeton, NJ: Princeton University Press.

Oberthür, Sebastian, and Olav Schram Stokke, eds. 2011. *Managing Institutional Complexity: Regime Interplay and Global Environmental Change*. Institutional Dimensions of Global Environmental Change. Cambridge, MA: MIT Press.

Ogden, Peter, and Howard Marano. 2016. "The First G-20 Fossil Fuel Subsidy Peer Review: Insights and Lessons from the United States and China." Washington, DC: Center for American Progress.

OGJ editors. 2020. "Norway Approves FID in Northern Lights CCS Project." *Oil and Gas Journal*, December 17. https://www.ogj.com/general-interest/article/14189278/norway-approves-fid-in-northern-lights-ccs-project.

Oil and Gas Climate Initiative. 2019. "Scaling up Action: Aiming for Net Zero Emissions." London: Oil and Gas Climate Initiative. https://oilandgasclimateinitiative.com/ogci-annual-reports/.

Oreskes, Naomi, and Erik M. Conway. 2011. *Merchants of Doubt: How a Handful of Scientists Obscured the Truth on Issues from Tobacco Smoke to Global Warming*. New York: Bloomsbury Press.

Ostrom, Elinor. 1990. *Governing the Commons: The Evolution of Institutions for Collective Action*. Cambridge: Cambridge University Press.

———. 2010. "Beyond Markets and States: Polycentric Governance of Complex Economic Systems." *American Economic Review* 100 (3): 641–72. https://doi.org/10.1257/aer.100.3.641.

Overdevest, Christine, and Jonathan Zeitlin. 2018. "Experimentalism in Transnational Forest Governance: Implementing European Union Forest Law Enforcement, Governance and Trade (FLEGT) Voluntary Partnership Agreements in Indonesia and Ghana: Transnational Forest Governance." *Regulation and Governance* 12 (1): 64–87. https://doi.org/10.1111/rego.12180.

Oye, Kenneth A., and James H. Maxwell. 1994. "Self-Interest and Environmental Management." *Journal of Theoretical Politics* 6 (4): 593–624.

Pacheco, Pablo, George Schoneveld, Ahmad Dermawan, Heru Komarudin, and Marcel Djama. 2018. "Governing Sustainable Palm Oil Supply: Disconnects, Complementarities, and Antagonisms between State Regulations and Private Standards: Governing Sustainable Palm Oil Supply." *Regulation and Governance* 14 (3): 568–98. https://doi.org/10.1111/rego.12220.

Parkinson, Giles. 2020. "Hornsdale Big Battery Doubles Savings to Consumers, and Keeps Lights On." *Renew Economy*, 28 February. https://reneweconomy.com.au/hornsdale-big-battery-doubles-savings-to-consumers-and-keeps-lights-on-85139/.

Parrillo, Nicholas R. 2017. "Federal Agency Guidance: An Institutional Perspective." Washington, DC: Administrative Conference of the United States.

———. 2019. "Federal Agency Guidance and the Power to Bind: An Empirical Study of Agencies and Industries." *Yale Journal on Regulation* 36 (1): 165–271.

Parson, Edward. 2003. *Protecting the Ozone Layer: Science and Strategy*. Oxford: Oxford University Press.

Penn, Ivan. 2020. "Its Electric Grid under Strain, California Turns to Batteries." *New York Times*, September 3. https://www.nytimes.com/2020/09/03/business/energy-environment/california-electricity-blackout-battery.html.

Picoli, Michelle C. A., Ana Rorato, Pedro Leitão, Gilberto Camara, Adeline Maciel, Patrick Hostert, and Ieda Del'Arco Sanches. 2020. "Impacts of Public and Private Sector Policies on Soybean and Pasture Expansion in Mato Grosso—Brazil from 2001 to 2017." *Land* 9 (1): 20. https://doi.org/10.3390/land9010020.

Piore, Michael J., and Charles F. Sabel. 1984. *The Second Industrial Divide: Possibilities for Prosperity*. New York: Basic Books.

Poikane, Sandra, Nikolaos Zampoukas, Angel Borja, Susan P. Davies, Wouter van de Bund, and Sebastian Birk. 2014. "Intercalibration of Aquatic Ecological Assessment Methods in the European Union: Lessons Learned and Way Forward." *Environmental Science and Policy* 44 (December): 237–46. https://doi.org/10.1016/j.envsci.2014.08.006.

"Povos Indígenas do Rio Negro Planejam o Futuro que Desejam." 2017. Instituto Socioambiental, June 14. https://www.socioambiental.org/pt-br/noticias-socioambientais/povos-indigenas-do-rio-negro-planejam-o-futuro-que-desejam.

Pyper, Julia. 2017. "Tesla, Greensmith, AES Deploy Aliso Canyon Battery Storage in Record Time." Greentech Media, January 31. https://www.greentechmedia.com/articles/read/aliso-canyon-emergency-batteries-officially-up-and-running-from-tesla-green.

———. 2019. "Tracking Progress on 100% Clean Energy Targets." Greentech Media, November 12. https://www.greentechmedia.com/articles/read/tracking-progress-on-100-clean-energy-targets.

Rabe, Barry G. 2004. *Statehouse and Greenhouse: The Emerging Politics of American Climate Change Policy*. Washington, DC: Brookings Institution Press.

Rada, Nicholas, Steven Helfand, and Marcelo Magalhães. 2019. "Agricultural Productivity Growth in Brazil: Large and Small Farms Excel." *Food Policy* 84 (April): 176–85. https://doi.org/10.1016/j.foodpol.2018.03.014.

Rajamani, Lavanya. 2013. "Differentiation in the Emerging Climate Regime." *Theoretical Inquiries in Law* 14 (1): 152–71.

Rajamani, Lavanya, and Daniel Bodansky. 2019. "The Paris Rulebook: Balancing Prescriptiveness with Flexibility." *International and Comparative Law Quarterly* 68 (4): 1023–40. https://doi.org/10.1017/S0020589319000320.

Raskolnikov, Alex. 2021. "Distributional Arguments, in Reverse." *Minnesota Law Review*. 105(4): 1583–1666.

Raustiala, Kal, and David G. Victor. 2004. "The Regime Complex for Plant Genetic Resources." *International Organization* 58 (2): 277–309. https://doi.org/10.1017/S0020818304582036.

Reed, Leslie H., Jr. 1997. "California Low-Emission Vehicle Program: Forcing Technology and Dealing Effectively with the Uncertainties." *Boston College Environmental Affairs Law Review* 24 (4): 695–793.

Rees, Joseph V. 1996. *Hostages of Each Other: The Transformation of Nuclear Safety since Three Mile Island*. Chicago: University of Chicago Press.

Revkin, Andrew C. 2009. "Scenes from a Climate Floor Fight." *Dot Earth* (*New York Times* blog). December 19. https://dotearth.blogs.nytimes.com/2009/12/19/scenes-from-a-climate-floor-fight.

Rhodes, Rod A. W. 1997. *Understanding Governance: Policy Networks, Governance, Reflexivity and Accountability*. Bristol, PA: Open University Press.

Riofrancos, Thea N. 2020. *Resource Radicals: From Petro-Nationalism to Post-Extractivism in Ecuador*. Radical Américas. Durham, NC: Duke University Press.

Riordan, Brendan. 2012. "Estimation of the Contribution of the Biosector to Ireland's Net Foreign Earnings: Methodology and Results." Paper presented at the annual conference of the Agricultural Economics Society of Ireland, October 18. http://www.aesi.ie/aesi2012/aesi2012riordan.pdf.

Roberts, David. 2018. "Solar Power's Greatest Challenge Was Discovered 10 Years Ago. It Looks like a Duck." *Vox*, August 29. https://www.vox.com/energy-and-environment/2018/3/20/17128478/solar-duck-curve-nrel-researcher.

Rocky Mountain Farmers Union et al. v. Corey, No. 12–15131. 2013. U.S. App. LEXIS 19258 (9th Cir.).

Rodrik, Dani. 2011. *The Globalization Paradox: Democracy and the Future of the World Economy*. New York: W. W. Norton.

Rodrik, Dani, and Charles F. Sabel. Forthcoming. "Building a Good Jobs Economy." In *A Political Economy of Justice*, edited by Danielle Allen, Yochai Benkler, Leah Downey, Rebecca Henderson, and Josh Simons. Chicago: University of Chicago Press. http://www2.law.columbia.edu/sabel/papers/Building%20a%20Good%20Jobs%20Economy%20November%202019_final.pdf.

Ronneberg, Espen. 2016. "Small Islands and the Big Issue: Climate Change and the Role of the Alliance of Small Island States." In *The Oxford Handbook of International Climate Change Law*, edited by Kevin R. Gray, Richard Tarasofsky, and Cinnamon Carlarne, 761–78. Oxford: Oxford University Press. https://doi.org/10.1093/law/9780199684601.003.0034.

Ropkins, Karl, and Angus J. Beck. 2000. "Evaluation of Worldwide Approaches to the Use of HACCP to Control Food Safety." *Trends in Food Science and Technology* 11 (1): 10–21. https://doi.org/10.1016/S0924-2244(00)00036-4.

Sabel, Charles F. 2019. "Sovereignty and Complex Interdependence: Some Surprising Indications of Their Compatibility." In *Ideas That Matter*, edited by Debra Satz and Annabelle Lever, 201–30. Oxford University Press. https://doi.org/10.1093/oso/9780190904951.003.0009.

Sabel, Charles F., Archon Fung, and Bradley Karkkainen. 1999. "Beyond Backyard Environmentalism." *Boston Review*, October 1. http://bostonreview.net/forum/charles-sabel-archon-fung-bradley-karkkainen-beyond-backyard-environmentalism.

Sabel, Charles F., and William H. Simon. 2011. "Minimalism and Experimentalism in the Administrative State." *Georgetown Law Journal* 100:53–93.

Sabel, Charles F., and David G. Victor. 2015. "Governing Global Problems under Uncertainty: Making Bottom-up Climate Policy Work." *Climatic Change* 144 (October): 15–27. https://doi.org/10.1007/s10584-015-1507-y.

———. 2016. "Making the Paris Process More Effective: A New Approach to Policy Coordination on Global Climate Change." Stanley Foundation. http://www.stanleyfoundation.org/publications/pab/Sabel-VictorPAB216.pdf.

Sabel, Charles F., and Jonathan Zeitlin. 1985. "Historical Alternatives to Mass Production: Politics, Markets and Technology in Nineteenth-Century Industrialization." *Past and Present* 108 (1): 133–76. https://doi.org/10.1093/past/108.1.133.

———. 2008. "Learning from Difference: The New Architecture of Experimentalist Governance in the EU." *European Law Journal* 14 (3): 271–327.

Sand, Peter H. 1991. "Lessons Learned in Global Environmental Governance." *Boston College Environmental Affairs Law Review* 18 (2): 213–77.

Santiago, Thaís Muniz Ottoni, Jill Caviglia-Harris, and José Luiz Pereira de Rezende. 2018. "Carrots, Sticks and the Brazilian Forest Code: The Promising Response of Small Landowners in the Amazon." *Journal of Forest Economics* 30 (January): 38–51. https://doi.org/10.1016/j.jfe.2017.12.001.

"SB-1078 Renewable Energy: California Renewables Portfolio Standard Program." 2002. California Legislative Information. https://leginfo.legislature.ca.gov/faces/billNavClient.xhtml?bill_id=200120020SB1078.

Scharpf, Fritz Wilhelm. 1997. *Games Real Actors Play: Actor-Centered Institutionalism in Policy Research*. Theoretical Lenses on Public Policy. Boulder, CO: Westview Press.

Schelling, Thomas C., ed. 1983. *Incentives for Environmental Protection*. MIT Press Series on the Regulation of Economic Activity 5. Cambridge, MA: MIT Press.

Schielein, Johannes, and Jan Börner. 2018. "Recent Transformations of Land-Use and Land-Cover Dynamics across Different Deforestation Frontiers in the Brazilian Amazon." *Land Use Policy* 76 (July): 81–94. https://doi.org/10.1016/j.landusepol.2018.04.052.

Schmalensee, Richard, Paul L. Joskow, A. Denny Ellerman, Juan Pablo Montero, and Elizabeth M. Bailey. 1998. "An Interim Evaluation of Sulfur Dioxide Emissions Trading." *Journal of Economic Perspectives* 12 (3): 53–68. https://doi.org/10.1257/jep.12.3.53.

Schmalensee, Richard, and Robert N. Stavins. 2013. "The SO_2 Allowance Trading System: The Ironic History of a Grand Policy Experiment." *Journal of Economic Perspectives* 27 (1): 103–22. https://doi.org/10.1257/jep.27.1.103.

———. 2017. "Lessons Learned from Three Decades of Experience with Cap and Trade." *Review of Environmental Economics and Policy* 11 (1): 59–79. https://doi.org/10.1093/reep/rew017.

Schmink, Marianne, Jeffrey Hoelle, Carlos Valério A. Gomes, and Gregory M. Thaler. 2019. "From Contested to 'Green' Frontiers in the Amazon? A Long-Term Analysis of São Félix do Xingu, Brazil." *Journal of Peasant Studies* 46 (2): 377–99. https://doi.org/10.1080/03066150.2017.1381841.

Schneider, Cecile, Emile Coudel, Federico Cammelli, and Philippe Sablayrolles. 2015. "Small-Scale Farmers' Needs to End Deforestation: Insights for REDD in São Felix do Xingu (Pará, Brazil)." *International Forestry Review* 17 (1): 124–42. https://doi.org/10.1505/146554815814668963.

Schoneveld, George C., Selma van der Haar, Dian Ekowati, Agus Andrianto, Heru Komarudin, Beni Okarda, Idsert Jelsma, and Pablo Pacheco. 2019. "Certification, Good Agricultural Practice and Smallholder Heterogeneity: Differentiated Pathways for Resolving Compliance Gaps in the Indonesian Oil Palm Sector." *Global Environmental Change* 57 (July): 101933. https://doi.org/10.1016/j.gloenvcha.2019.101933.

Schwartzman, Stephan. 1991. "Deforestation and Popular Resistance in Acre: From Local Social Movement to Global Network." *Centennial Review* 35 (2): 397–422.

Schwartzman, Stephan, Ane Alencar, Hilary Zarin, and Ana Paula Santos Souza. 2010. "Social Movements and Large-Scale Tropical Forest Protection on the Amazon Frontier: Conservation from Chaos." *Journal of Environment and Development* 19 (3): 274–99. https://doi.org/10.1177/1070496510367627.

Scott, Joanne, and Jane Holder. 2006. "Law and New Environmental Governance in the European Union." In *Law and New Governance in the EU and the US*, edited by Gráinne de Búrca and Joanne Scott, 211–42. Oxford: Hart Publishing.

Seymour, Frances, Leony Aurora, and Joko Arif. 2020. "The Jurisdictional Approach in Indonesia: Incentives, Actions, and Facilitating Connections." *Frontiers in Forests and Global Change* 3 (November): 124. https://doi.org/10.3389/ffgc.2020.503326.

Seymour, Frances J., and Nancy L. Harris. 2019. "Reducing Tropical Deforestation." *Science* 365 (6455): 756–57. https://doi.org/10.1126/science.aax8546.

Shortle, Ger, and Phil Jordan, eds. 2017. *Agricultural Catchments Programme: Phase 2 Report.* Wexford, Ireland: Teagasc.

Sivaram, Varun. 2018. *Taming the Sun: Innovations to Harness Solar Energy and Power the Planet.* Cambridge, MA: MIT Press.

Sivaram, Varun, Julio Friedmann, and Colin Cunliff. 2020. "To Confront the Climate Crisis, the US Should Launch a National Energy Innovation Mission." New York: Columbia Center on Global Policy. https://www.energypolicy.columbia.edu/research/op-ed/confront-climate-crisis-us-should-launch-national-energy-innovation-mission.

Skidmore, Marin. 2020. "*Agriculture, Environmental Policy, and Climate: Essays on Cattle Ranching in the Brazilian Amazon.*" PhD diss., University of Wisconsin at Madison.

Skjærseth, Jon Birger. 1998. "The Making and Implementation of North Sea Commitments: The Politics of Environmental Participation." In *The Implementation and Effectiveness of International Environmental Commitments: Theory and Practice*, edited by David G. Victor, Kal Raustiala, and Eugene B. Skolnikoff, 327–76. Cambridge, MA: MIT Press.

"Smallholder Crucial to Preserving the Amazon Rainforest." 2019. Solidaridad, January 29. https://www.solidaridadnetwork.org/news/smallholders-crucial-to-preserving-the-amazon-rainforest/.

Smith, Abby. 2019. "States Fight for Their Right to Follow California Car Rules." *Bloomberg Law*, January 11. https://news.bloomberglaw.com/environment-and-energy/states-fight-for-their-right-to-follow-california-car-rules.

Smith, Adam. 1976. *The Wealth of Nations.* Edited by Edwin Canaan. Chicago: University of Chicago Press.

Social Learning Group. 2001. *Learning to Manage Global Environmental Risks, Vol. 1: A Comparative History of Social Responses to Climate Change, Ozone Depletion and Acid Rain.* Politics, Science, and the Environment. Cambridge, MA: MIT Press.

Soekkha, Hans M., ed. 1997. *Aviation Safety: Human Factors—System Engineering Flight Operations—Economics Strategies—Management.* London: CRC Press. https://doi.org/10.1201/9780429070372.

Solomon, Susan, Rolando R. Garcia, F. Sherwood Rowland, and Donald J. Wuebbles. 1986. "On the Depletion of Antarctic Ozone." *Nature* 321 (6072): 755–58. https://doi.org/10.1038/321755a0.

St. John, Jeff. 2021. "California Sets $200M Budget for 'Complex, Multi-Property Microgrid' Projects." Greentech Media, January 15. https://www.greentechmedia.com/articles/read/california-sets-200m-budget-for-complex-multi-property-microgrid-projects.

Stavins, Robert. 1988. "Project 88—Harnessing Market Forces to Protect Our Environment: Initiatives for the New President. A Public Policy Study Sponsored by Senator Timothy E. Wirth, Colorado, and Senator John Heinz, Pennsylvania." 1988.

———. 2019. "The Madrid Climate Conference's Real Failure Was Not Getting a Broad Deal on Global Carbon Markets." *Conversation*, December 18. https://theconversation.com/the-madrid-climate-conferences-real-failure-was-not-getting-a-broad-deal-on-global-carbon-markets-129001.

Stewart, Richard B., and Jonathan Wiener. 2004. "Practical Climate Change Policy." *Issues in Science and Technology* 20 (2): 71–78.

Stickler, Claudia, Amy Duchelle, Juan Pablo Ardila, Daniel Nepstad, Olivia David, Charlotta Chan, Juan Rojas, et al. 2018. "The State of Jurisdictional Sustainability: Synthesis for Practitioners and Policymakers." San Francisco: Earth Innovation Institute. https://earthinnovation.org/state-of-jurisdictional-sustainability/.

Stokes, Leah C. 2020. *Short Circuiting Policy: Interest Groups and the Battle over Clean Energy and Climate Policy in the American States*. Studies in Postwar American Political Development. New York: Oxford University Press.

Stone, Jon. 2018. "EU to Refuse to Sign Trade Deals with Countries That Don't Ratify Paris Climate Change Accord." *Independent*, February 12. https://www.independent.co.uk/news/world/europe/eu-trade-deal-paris-climate-change-accord-agreement-cecilia-malmstrom-a8206806.html.

Streck, Charlotte, and Jolene Lin. 2010. "Making Markets Work—A Review of CDM Performance and the Need for Reform." In *Crucial Issues in Climate Change and the Kyoto Protocol: Asia and the World*, edited by Kheng Lian Koh, Lin Heng Lye, Jolene Lin, and World Scientific, 181–232. Singapore: World Scientific Publishing Co. http://public.eblib.com/choice/publicfullrecord.aspx?p=1681232.

Sunstein, Cass R., and Reid Hastie. 2015. *Wiser: Getting beyond Groupthink to Make Groups Smarter*. Boston: Harvard Business Review Press.

Sustainable Development Solutions Network. 2020. "Zero Carbon Action Plan." New York: Sustainable Development Solutions Network. https://irp-cdn.multiscreensite.com/6f2c9f57/files/uploaded/zero-carbon-action-plan%20%281%29.pdf.

Taylor, Margaret R., Edward S. Rubin, and David A. Hounshell. 2003. "Effect of Government Actions on Technological Innovation for SO_2 Control." *Environmental Science and Technology* 37 (20): 4527–34.

Thaler, Gregory M., Cecilia Viana, and Fabiano Toni. 2019. "From Frontier Governance to Governance Frontier: The Political Geography of Brazil's Amazon Transition." *World Development* 114 (February): 59–72. https://doi.org/10.1016/j.worlddev.2018.09.022.

Thorne, Fiona, Patrick R. Gillespie, Trevor Donnelan, Kevin Hanrahan, Anne Kinsella, and Doris Läpple. 2017. *The Competitiveness of Irish Agriculture*. Carlow, Ireland: Teagasc. https://www.teagasc.ie/media/website/publications/2017/The-Competitiveness-of-Irish-Agriculture.pdf.

Tolba, Mostafa Kamal, and Iwona Rummel-Bulska. 1998. *Global Environmental Diplomacy: Negotiating Environmental Agreements for the World, 1973–1992*. Global Environmental Accord. Cambridge, MA: MIT Press.

Trebilcock, Bob. 2017. "How They Did It: Supplier Trust at General Motors." *Supply Chain Management Review* (May–June): 20–25.

Turner, James Morton, and Andrew C. Isenberg. 2018. *The Republican Reversal: Conservatives and the Environment from Nixon to Trump*. Cambridge, MA: Harvard University Press.

United Nations Environment Programme (UNEP). 1985. "Vienna Convention for the Protection of the Ozone Layer." Nairobi: United Nations Environment Programme.

———. 1987. "Montreal Protocol on Substances That Deplete the Ozone Layer." Nairobi: Ozone Secretariat of the United Nations Environment Programme.

———. 1989a. "Decision I/3: Establishment of Assessment Panels." United Nations Environment Programme. https://ozone.unep.org/treaties/montreal-protocol/meetings/first-meeting-parties/decisions/decision-i3-establishment-assessment-panels.

———. 1989b. "UNEP/OzL.Pro.1/5—Report of the First Meeting of the Parties to the Montreal Protocol." United Nations Environment Programme. https://ozone.unep.org/Meeting_Documents/mop/01mop/MOP_1.shtml.

———. 2020. "Oil and Gas Methane Partnership (OGMP) 2.0 Framework." Nairobi: United Nations Environment Programme. https://www.ccacoalition.org/en/resources/oil-and-gas-methane-partnership-ogmp-20-framework.

United Nations Framework Convention on Climate Change (UNFCCC). 1992. "United Nations Framework Convention on Climate Change." Bonn, Germany: United Nations. http://unfccc.int/essential_background/convention/background/items/1350.php.

———. 1995. "The Berlin Mandate." FCCC/CP/1995/7/Add.1. Berlin: United Nations. https://unfccc.int/resource/docs/cop1/07a01.pdf.

———. 2015. "The Paris Agreement." FCCC/CP/2015/L.9/Rev1. Katowice, Poland: United Nations Framework Convention on Climate Change. https://unfccc.int/process/conferences/pastconferences/paris-climate-change-conference-november-2015/paris-agreement.

———. 2018. "Report of the Conference of the Parties Serving as the Meeting of the Parties to the Paris Agreement on the Third Part of Its First Session, Held in Katowice from 2 to 15 December 2018." UN Doc. FCCC/PA/CMA/2018/3. Katowice, Poland: United Nations Framework Convention on Climate Change. https://undocs.org/pdf?symbol=en/FCCC/PA/CMA/2018/3.

———. 2019. "UNFCCC Technical Expert Meetings 2019: Summary for Policymakers." No. 978-92-9219-185-6. Bonn: United Nations Climate Change Secretariat.

———. 2020. "Yearbook of Global Climate Action 2020: Marrakech Partnership for Global Climate Action." Bonn: United Nations Framework Convention on Climate Change. https://unfccc.int/documents/267246.

———. 2021. "UNCOP26 Campaigns." United Nations Climate Change Conference UK 2021. https://ukcop26.org/uk-presidency/campaigns/.

United Nations General Assembly (UNGA). 1988. "Protection of Global Climate for Present and Future Generations of Mankind: Resolution / Adopted by the General Assembly." A/RES/43/53, December 6. http://www.un.org/documents/ga/res/43/a43r053.htm.

———. 1990. "Protection of Global Climate for Present and Future Generations of Mankind: Resolution / Adopted by the General Assembly." A/RES/45/212, December 21. http://www.un.org/documents/ga/res/45/a45r212.htm.

United Nations Industrial Development Organization (UNIDO). 2015. "Toolkit for Sustainable Compliance with the Methyl Bromide Phase-Out." https://www.unido.org/sites/default/files/2015-11/UNIDO_TOOLKIT_for_sustainable_MB_phase-out_FIN__2__0.pdf.

United States. 2015. "United States Self-Review of Fossil Fuel Subsidies." Paris: Organization for Economic Cooperation and Development. http://www.oecd.org/fossil-fuels/publication/United%20States%20Self%20review%20USA%20FFSR%20Self-Report%202015%20FINAL.pdf.

"Update of Norway's Nationally Determined Contribution." 2020. United Nations Framework Convention on Climate Change. https://www4.unfccc.int/sites/ndcstaging/PublishedDocuments/Norway%20First/Norway_updatedNDC_2020%20(Updated%20submission).pdf.

US Congress. 1977. Clean Air Act Amendments of 1977, Pub. L. 95–95, 91 Stat. 685. https://www.govinfo.gov/content/pkg/STATUTE-91/pdf/STATUTE-91-Pg685.pdf.

Vargas, Daniel. 2020. "*Amazônia: O Desafio do Desenvolvimento.*" FGV-Escola de Economia São Paulo. https://open.spotify.com/episode/649MKepddseslXglegFaDI.

Varns, Theodore, Rane Cortez, Lex Hovani, and Paul Kingsbury. 2018. "São Félix Do Xingu, Brazil: A Jurisdictional Approach to Conserving the Amazon." Arlington, VA: Nature Conservancy. https://www.nature.org/content/dam/tnc/nature/en/documents/TNC_JurisdictionalApproaches_CaseStudies_Brazil.pdf.

Vermeule, Adrian. 2016. *Law's Abnegation: From Law's Empire to the Administrative State.* Cambridge, MA: Harvard University Press.

Victor, David G. 2001. *The Collapse of the Kyoto Protocol and the Struggle to Slow Global Warming.* Princeton, NJ: Princeton University Press.

———. 2011. *Global Warming Gridlock: Creating More Effective Strategies for Protecting the Planet.* Cambridge: Cambridge University Press.

———. 2015. "Why Paris Worked: A Different Approach to Climate Diplomacy." *Yale Environment 360*, December 15. http://e360.yale.edu/features/why_paris_worked_a_different_approach_to_climate_diplomacy.

———. 2017. "Jump-Starting and Accelerating the Implementation of the Paris Agreement." Brookings, August 31. https://www.brookings.edu/blog/planetpolicy/2017/08/31/jump-starting-and-accelerating-the-implementation-of-the-paris-agreement/.

———. 2019. "We Have Climate Leaders. Now We Need Followers." *New York Times*, December 13. https://www.nytimes.com/2019/12/13/opinion/climate-change-madrid.html.

Victor, David G., and Lesley A. Coben. 2005. "A Herd Mentality in the Design of International Environmental Agreements?" *Global Environmental Politics* 5 (1): 24–57.

Victor, David G., Sadie Frank, and Eric Gesick. 2020. "Making Climate Policy Stick." Washington, DC: Brookings Institution. https://www.brookings.edu/blog/planetpolicy/2020/12/09/making-climate-policy-stick/.

Victor, David G., Frank W. Geels, and Simon Sharpe. 2019. "Accelerating the Low Carbon Transition." Washington, DC: Brookings Institution. https://www.brookings.edu/research/accelerating-the-low-carbon-transition/.

Victor, David G., and Bruce Jones. 2018. "Undiplomatic Action: A Practical Guide to the New Politics and Geopolitics of Climate Change." Washington, DC: Brookings Institution. https://www.brookings.edu/research/undiplomatic-action-a-practical-guide-to-the-new-politics-and-geopolitics-of-climate-change/.

Victor, David G., and Julian E. Salt. 1995. "Keeping the Climate Treaty Relevant." *Nature* 373 (6512): 280–82. https://doi.org/10.1038/373280a0.

Victor, David G., Dadi Zhou, Essam Hassan Mohamed Ahmed, Pradeep Kumar Dadhich, Jos Olivier, H-Holger Rogner, Kamel Sheikho, and Mitsutsune Yamaguchi. 2014. "Introductory Chapter." In *Climate Change 2014 Mitigation of Climate Change: Working Group III Contribution to the Fifth Assessment Report of the Intergovernmental Panel on Climate Change*, edited by Ottmar Edenhofer, Ramón Pichs-Madruga, Youba Sokona, Ellie Farahani, Susanne Kadner, Kristin Seyboth, Anna Adler, et al., 114–43. Cambridge: Cambridge University Press. https://doi.org/10.1017/CBO9781107415416.

Vogel, David. 1997. *Trading Up: Consumer and Environmental Regulation in a Global Economy.* Cambridge, MA: Harvard University Press.

———. 2018. *California Greenin': How the Golden State Became an Environmental Leader.* Princeton, NJ: Princeton University Press.

Vogler, John. 2007. "The International Politics of Sustainable Development." In *Handbook of Sustainable Development*, edited by Giles Atkinson, Simon Dietz, and Eric Neumayer, 430–46. Cheltenham, UK: Edward Elgar Publishing. https://doi.org/10.4337/9781847205223.

Voigt, Christina, and Felipe Ferreira. 2016. "Differentiation in the Paris Agreement." *Climate Law* 6 (1–2): 58–74. https://doi.org/10.1163/18786561-00601004.

von Hippel, Eric. 2006. *Democratizing Innovation*. Cambridge, MA: MIT Press.

Voulvoulis, Nikolaos, Karl Dominic Arpon, and Theodoros Giakoumis. 2017. "The EU Water Framework Directive: From Great Expectations to Problems with Implementation." *Science of the Total Environment* 575:358–66.

Wallace, Richard H., Carlos Valério A. Gomes, and Natalie A. Cooper. 2018. "The Chico Mendes Extractive Reserve: Trajectories of Agro-Extractive Development in Amazonia." *Desenvolvimento e Meio Ambiente* 48 (November). https://doi.org/10.5380/dma.v48i0.58836.

Ward, Allen, Jeffrey K. Liker, John J. Cristiano, and Durward K. Sobek II. 1995. "The Second Toyota Paradox: How Delaying Decisions Can Make Better Cars Faster." *Sloan Management Review* 36:43–61.

"WFD Guidance Documents." n.d. European Commission. http://ec.europa.eu/environment/water/water-framework/facts_figures/guidance_docs_en.htm.

Weiss, Edith Brown. 1993. "International Environmental Law: Contemporary Issues and the Emergence of a New World Order." *Georgetown Law Journal* 81 (3): 675–710.

West, Thales A. P., and Philip M. Fearnside. 2021. "Brazil's Conservation Reform and the Reduction of Deforestation in Amazonia." *Land Use Policy* 100 (105072). https://doi.org/10.1016/j.landusepol.2020.105072.

Western, David. 2020. *We Alone: How Humans Have Conquered the Planet and Can Also Save It*. New Haven, CT: Yale University Press.

Whately, Marussia, Maura Campanili, and Adalberto Veríssimo, eds. 2013. *Programa Municípios Verdes: Lições Aprendidas e Desafios Para 2013/2014*. Belém: Governo do Estado do Pará/Programa Municípios.

Womack, James P., Daniel T. Jones, and Daniel Roos. 2007. *The Machine That Changed the World: The Story of Lean Production—Toyota's Secret Weapon in the Global Car Wars That Is Revolutionizing World Industry*. London: Free Press.

Wong, Anna (CARB emissions compliance staff, advanced clean cars branch). 2015. Interview with Lauren Packard.

World Trade Organization (WTO). 2001a. "European Communities—Measures Affecting Asbestos and Asbestos-Containing Products (AB2000–11)." WT/DS135/AB/R. Geneva: World Trade Organization.

———. 2001b. "United States—Import Prohibition of Certain Shrimp and Shrimp Products." DS58. Geneva: World Trade Organization.

Yellen, Janet. 1998. "Testimony of Dr. Janet Yellen, Chair, Council of Economic Advisers before the House Commerce Committee on the Economics of the Kyoto Protocol." March 4. https://clintonwhitehouse2.archives.gov/WH/EOP/CEA/html/19980304.html.

Yi, Yuan. 2020. "Malfunctioning Machinery: The Global Making of Chinese Cotton Mills, 1877–1937." https://doi.org/10.7916/D8-NAM9-5835.

Young, Oran R. 2017. "Conceptualization: Goal Setting as a Strategy for Earth System Governance." In *Governing through Goals*, edited by Norichika Kanie and Frank Biermann, 31–52. Cambridge, MA: MIT Press. https://doi.org/10.7551/mitpress/10894.003.0007.

Zhao, Jimin, and Leonard Ortolano. 2003. "The Chinese Government's Role in Implementing Multilateral Environmental Agreements: The Case of the Montreal Protocol." *China Quarterly* 175 (September): 708–25. https://doi.org/10.1017/S0305741003000419.

INDEX

A NOTE ON THE TYPE

This book has been composed in Adobe Text and Gotham. Adobe Text, designed by Robert Slimbach for Adobe, bridges the gap between fifteenth- and sixteenth-century calligraphic and eighteenth-century Modern styles. Gotham, inspired by New York street signs, was designed by Tobias Frere-Jones for Hoefler & Co.